Lóegaire Humphrey (Ed.)

Albion Park Railway Station

Lóegaire Humphrey (Ed.)

Albion Park Railway Station

CityRail, South Coast railway line, New South Wales, Albion Park Rail, New South Wales, Albion Park, New South Wales

Claud Press

Imprint

Permission is granted to copy, distribute and/or modify this document under the terms of the GNU Free Documentation License, Version 1.2 or any later version published by the Free Software Foundation; with no Invariant Sections, with the Front-Cover Texts, and with the Back- Cover Texts. A copy of the license is included in the section entitled "GNU Free Documentation License".

All parts of this book are extracted from Wikipedia, the free encyclopedia (www.wikipedia.org).

You can get detailed informations about the authors of this collection of articles at the end of this book. The editors (Ed.) of this book are no authors. They have not modified or extended the original texts.

Pictures published in this book can be under different licences than the GNU Free Documentation License. You can get detailed informations about the authors and licences of pictures at the end of this book.

The content of this book was generated collaboratively by volunteers. Please be advised that nothing found here has necessarily been reviewed by people with the expertise required to provide you with complete, accurate or reliable information. Some information in this book maybe misleading or wrong. The Publisher does not guarantee the validity of the information found here. If you need specific advice (f.e. in fields of medical, legal, financial, or risk management questions) please contact a professional who is licensed or knowledgeable in that area.

Any brand names and product names mentioned in this book are subject to trademark, brand or patent protection and are trademarks or registered trademarks of their respective holders. The use of brand names, product names, common names, trade names, product descriptions etc. even without a particular marking in this works is in no way to be construed to mean that such names may be regarded as unrestricted in respect of trademark and brand protection legislation and could thus be used by anyone.

Cover image: www.ingimage.com
Concerning the licence of the cover image please contact ingimage.

Publisher:
Claud Press is a trademark of
International Book Market Service Ltd., 17 Rue Meldrum, Beau Bassin, 1713-01 Mauritius
Email: info@bookmarketservice.com
Website: www.bookmarketservice.com

Published in 2011

Printed in: U.S.A., U.K., Germany. This book was not produced in Mauritius.

ISBN: 978-613-6-52040-7

Contents

Albion Park railway station

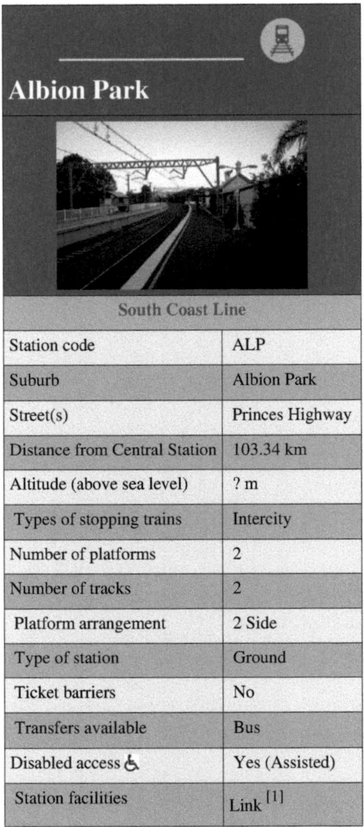

Station code	ALP
Suburb	Albion Park
Street(s)	Princes Highway
Distance from Central Station	103.34 km
Altitude (above sea level)	? m
Types of stopping trains	Intercity
Number of platforms	2
Number of tracks	2
Platform arrangement	2 Side
Type of station	Ground
Ticket barriers	No
Transfers available	Bus
Disabled access &	Yes (Assisted)
Station facilities	Link [1]

Albion Park railway station is on the Cityrail South Coast line in New South Wales, serving the township of Albion Park Rail. This township is east of the town of Albion Park proper. The station opened in 1887 (as Oak Flats),[2] and a second platform was opened in 2001 as part of the extension of electrification to Kiama.[3] A station 3 km north of the present station opened in 1887 as 'Albion Park', but was later renamed 'Yallah'[4] and closed in 1974.[5]

Albion Park has approximately one service to Sydney Terminal and one to Kiama every hour, during weekdays. Passengers heading to Nowra must chage at Kiama for a shuttle service to Nowra, however some peak hour services continue to Nowra.

Platforms and services

Platform	Line	Stopping Pattern	Notes and Comments
①	South Coast Line	Intercity services to Dapto, Wollongong and Sydney Terminal	
②	South Coast Line	Intercity services to Kiama; Peak Intercity services to Bomaderry	

Transport links

Premier Illawarra runs seven routes to and fron Albion Park railway station:

- **37** - Wollongong Loop
- **43** - Between Dapto and Port Kembla
- **51** - to University of Wollongong
- **54** - to Wollongong station
- **57** - Wollongong Loop
- **70** - to Shellharbour District
- **76** - to Shellharbour District

Neighbouring stations

Preceding station	CityRail	Following station
Oak Flats *towards Bomaderry (Nowra)*	South Coast Line	Dapto *towards Central*

References

[1] http://www.cityrail.info/facilities/facilities.jsp?n=3&giveOutput=true&facility=

[2] "Albion Park railway station" (http://www.nswrail.net/locations/show.php?name=NSW:Albion+Park&line=NSW:south_coast:0). www.nswrail.net. . Retrieved 2007-06-26.

[3] Gullick, J. *Kiama Electrification- some facts and figures*. Railway Digest, January 2002.

[4] Wollongong City Library Yallah History (http://www.wollongong.nsw.gov.au/library/localinfo/yallah/history.html), accessed 10 July 2007

[5] "Yallah railway station" (http://www.nswrail.net/locations/show.php?name=NSW:Yallah&line=NSW:south_coast:0). www.nswrail.net. . Retrieved 2007-07-10.

CityRail

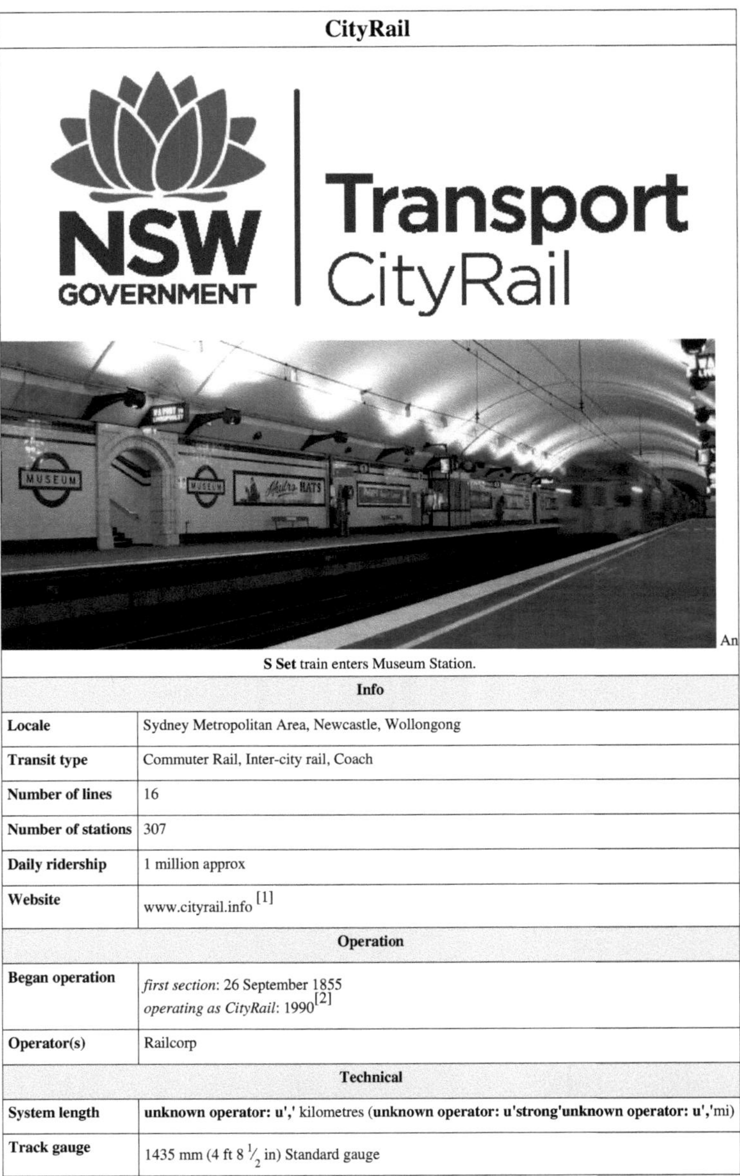

<table>
<tr><td colspan="2" align="center">CityRail</td></tr>
<tr><td colspan="2" align="center">S Set train enters Museum Station.</td></tr>
<tr><td colspan="2" align="center">Info</td></tr>
<tr><td>Locale</td><td>Sydney Metropolitan Area, Newcastle, Wollongong</td></tr>
<tr><td>Transit type</td><td>Commuter Rail, Inter-city rail, Coach</td></tr>
<tr><td>Number of lines</td><td>16</td></tr>
<tr><td>Number of stations</td><td>307</td></tr>
<tr><td>Daily ridership</td><td>1 million approx</td></tr>
<tr><td>Website</td><td>www.cityrail.info [1]</td></tr>
<tr><td colspan="2" align="center">Operation</td></tr>
<tr><td>Began operation</td><td>first section: 26 September 1855
operating as CityRail: 1990[2]</td></tr>
<tr><td>Operator(s)</td><td>Railcorp</td></tr>
<tr><td colspan="2" align="center">Technical</td></tr>
<tr><td>System length</td><td>unknown operator: u',' kilometres (unknown operator: u'strong'unknown operator: u','mi)</td></tr>
<tr><td>Track gauge</td><td>1435 mm (4 ft 8 $\frac{1}{2}$ in) Standard gauge</td></tr>
</table>

CityRail is an operating brand of RailCorp, a corporation owned by the state government of New South Wales, Australia. It is responsible for providing commuter rail services, and some coach services, in and around Sydney, Newcastle and Wollongong, the three largest cities of New South Wales. It is also the name of the network on which the services run. It is part of the MyZone ticketing system.

Construction of what is now the CityRail network began 3 July 1850. Today it consists of 305 stations and over 2060 km (1280 mi) of track, extending to the upper Hunter Valley and the Shoalhaven area. Four new lines are now in various stages of planning and construction. CityRail and the state's Transport Construction Authority are currently engaged in a process of "sectorisation", a project called "Rail Clearways", in an effort to reduce its operational complexity.

Established under the *Transport Administration Act (NSW) 1988* around 1990, CityRail is a "product group" of Rail Corporation New South Wales (**Railcorp**), the state-owned corporation (SOC) that operates the New South Wales railways. It is a sister group of CountryLink, which operates rail and coach services in regional New South Wales.

Most of the CityRail system is electrified with 1500 V DC supplied by overhead wire; some isolated sections outside the Sydney metropolitan area are operated by single-deck diesel railcars. All CityRail electric trains are double-deck multiple units.

An average of one million trips is made to and from 307 stations each weekday.[3]

Operations

Fleet

In 2009 CityRail ran ten types of rolling stock, in two categories: electric multiple units (EMUs) for suburban and interurban working, and diesel multiple units (DMUs) for interurban and regional lines running through less populated areas. All CityRail electric trains use 1500 V DC overhead electrification and travel on 1435 mm standard gauge tracks.The new waratah trains need more energy. All electric rolling stock has been double deck since the early 1990s.

The CityRail network is divided into three sectors, based around three maintenance depots.[4] EMU trainsets are identified by target plates, which are exhibited on the front lower nearside of driving carriages.[5] Target designations and set numbers are used in identifying EMU trainsets. The composition and formations of trainsets, and the target designations

The concourse of Central Railway Station, the main station on the CityRail network. The station opened in its present location in 1906.

are subject to alteration. The target designation originally identified the depot at which a trainset was based, e.g. "M" for Mortdale, "F" for Flemington, "H" for Hornsby and "B" for Punchbowl (on the Bankstown line). However, the introduction of a variety of EMU types led to the target designation's use as a means of identifying the type of trainset, more like a vehicle or locomotive number. Hence, "M" is now used for the Millennium trainsets.

A Waratah Train, the newest train in CityRail's
fleet awaiting departure at Campbelltown, on a
service to Circular Quay via the South Line

CityRail *"Millennium trains"* operate on the
Airport & East Hills, South, Bankstown, Inner
West Lines to and from the City Circle

Cityrail maintenance sectors

Sector #	Depot	Serviced lines	Target plate
1	Mortdale	Illawarra and Eastern Suburbs, South Coast	Red
2	Flemington	Cumberland, Airport and East Hills, Olympic Park Sprint, Carlingford, South, Bankstown	Blue
3	Hornsby	North Shore, Northern, Western, Richmond, Newcastle, Blue Mountains, Central Coast	Black

All double deck InterUrban (DDIU or **V set**) EMU trains, which operate on the Blue Mountains, Newcastle and Central Coast, and South Coast lines, are serviced at Flemington Depot, and all M set and H set trains, which have a green target plate, are serviced at Eveleigh Maintenance Centre near Redfern station.

In November 2006 the NSW government and RailCorp/CityRail announced the manufacturer for the new trains (626 carriages) set to replace the L, R & S sets. The chosen manufacturer RelianceRail/Downer EDI is the same company that built the M (Millennium) sets. This has been said to be the largest order of rolling stock in Australia's history, at a cost of \$3.6 billion, which includes a brand new maintenance facility at Auburn. The new trains are named the 'Waratah' which is the NSW Government State's floral emblem. The first 4 carriage prototype vehicle began network testing by April 2010 as it's design was complete in July 2009 and delivered to Newcastle for manufacturing by late 2009. At July 2010 only the pre-production test set had been delivered and began network testing in August 2010 and then by November 2010, three sets were undergoing network testing. The first of these Waratah trains were scheduled to enter passenger service between December 2010 and April 2011 with 78 train sets delivered by 2013, however another delay due to asbestos contamination has further delayed its entry into service from June 2011, and fulfillment of orders scheduled for late 2013 to 2014.

Ticketing

CityRail's current automated ticketing system, dating from 1992, is based on magnetic stripe technology and is interoperable with the government's buses and ferries.

Unlike the ticketing systems of other cities in Australia, most of CityRail's ticket prices are calculated on the distance travelled and are inexpensive by world standards.[6]

Entry to privately owned train stations at Sydney Airport requires a Station Access Fee in addition to the train fare.[7]

CityRail Mytrain ticket 2010

Tcard

Tcard is the name of the contactless smartcard ticketing system that is planned to be introduced on public transport in Sydney, New South Wales by 2012. The Smart Card is expected to replace the existing Automated Fare Collection System on CityRail services. On 12 April 2010, the NSW Government re-announced that a new contract had been awarded to the PeaRl Consortium for the roll-out of the Tcard system, after the previous contract with ERG collapsed in 2008. The project is 13 years behind schedule, and was planned to have completed before the commencement of the Sydney Olympic Games in 2000. Limited progress has developed since then, however a selected few stations in 2010 have had place-holders for Tcard validators installed near station entrances/exits in preparation for the Tcard's long awaited arrival.

Performance

According to the 2003 'Parry report', "The interaction of metropolitan, suburban, intercity and freight lines and services has resulted in an overly complex system."[6] This complexity has contributed in part to the organisation being widely criticised for poor reliability and safety. CityRail is also enormously expensive. RailCorp requires a government subsidy of close to \$1.8 billion a year, approximately 5% of the state budget and more than three times what it collects in fares. "There is an overwhelming sense," the report concluded, "that CityRail does not promote a real commitment to quality, customer focus and a service culture."

On-time running has improved since new timetables were introduced in 2005 and 2006. This is because the travelling times between stations were extended in the timetable. In April 2008, 99.6% of all services ran, and 92.6% of these services arrived within five minutes of their scheduled arrival time.[8] However a 2007 report by Hong Kong's Mass Transit Railway Corporation found that Sydney's train system reliability levels lagged behind international benchmarks.[9]

Network

In 2009 CityRail operated 11 suburban lines, four intercity lines, and one regional line. The standard network map is shown here. [10]

Suburban lines

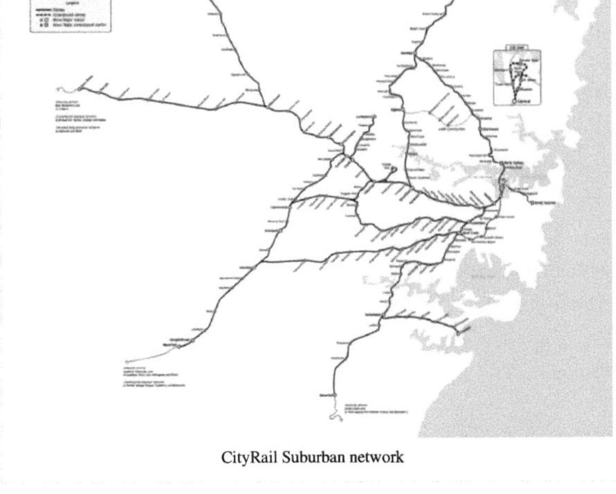

CityRail Suburban network

- **Eastern Suburbs & Illawarra Line** - Between Bondi Junction and Waterfall/Cronulla.
- **Bankstown Line** - Between Central and Liverpool/Lidcombe, via City Circle (clockwise) and Bankstown.
- **Inner West Line** - Between Central and Bankstown/Liverpool, via City Circle (anticlockwise) and Strathfield.
- **Airport & East Hills Line** - Between Central and Macarthur, via City Circle (clockwise) and Sydenham (peak) or Wolli Creek.
- **South Line** - Between Central and Campbelltown, via City Circle (anticlockwise) and Granville.
- **Cumberland Line** - Between Blacktown and Campbelltown.
- **Western Line** - Between Central and Emu Plains or Richmond. This includes the:

  - **Richmond Line**.
- **North Shore Line** - Between Central and Berowra (via Chatswood).
- **Carlingford Line** - Between Clyde and Carlingford.
- **Olympic Park Sprint** - Between Lidcombe and Olympic Park, extending to Central in off-peak and during special events.
- **Northern Line** - Between Epping and Hornsby (via Macquarie Park, City and Strathfield).

* In peak hour on the North Shore line, some outer-suburban services run to Gosford and Wyong, and some Western Line services extend to Springwood.

In peak hour, some City Circle services run in the opposite direction than normal

Intercity services

Intercity lines are shown in grey on CityRail maps, with the line colour on the stations.

- **South Coast Line** - Between Central** and Bomaderry (Nowra) or Port Kembla.
- **Southern Highlands Line** - Between Campbelltown** and Goulburn.
- **Blue Mountains Line** - Between Central** and Lithgow.
- **Newcastle & Central Coast Line** - Between Central and Newcastle.

** Some peak services and most weekend services on the South Coast Line run to/from Bondi Junction, some on the Southern Highlands Line to/from Central, and some on the Blue Mountains Line to/from Hornsby. Southern Highlands services run to Central only in the morning and from Central in the afternoon and evening. At other times, a change of train is required at Campbelltown or Macarthur.

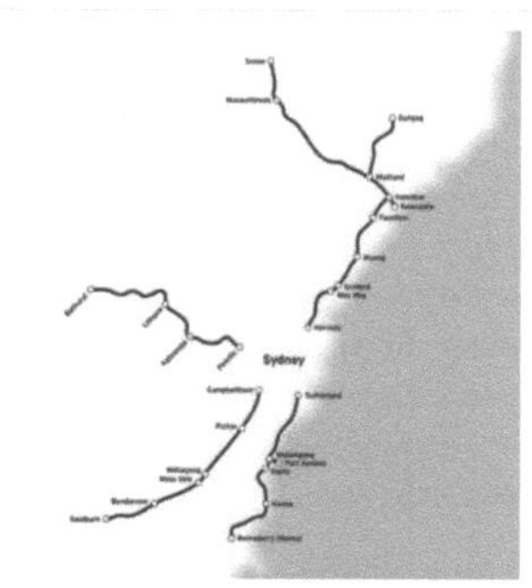

CityRail intercity and regional network as of 2000.

Regional services

- **Hunter Lines** - Between Newcastle and Telarah, with less frequent services to Dungog or Scone.

Connecting bus services

CityRail operates several bus routes along corridors where the railway line has been closed to passengers. These bus services appear in CityRail timetables and accept CityRail tickets, but they are operated by private-sector bus companies contracted by CityRail. In 2006 these CityRail bus services are:

- **Bowral to Picton Loop Line** - Bowral to Picton via Thirlmere. weekdays only
- **South Coast to Southern Highlands Line** - Bundanoon/Bowral to Wollongong via Robertson.
- **Lithgow to Bathurst** - Lithgow to Bathurst via Mt Lambie.
- **Fassifern to Toronto** - Fassifern to Toronto via Blackalls Park.

NightRide

To provide a passenger service between midnight and 5.00 am while leaving the tracks clear of trains for maintenance work, a parallel bus service was established in 1989. The **NightRide** operates typically at hourly intervals (some routes depart more frequently on weekends). NightRide services are run by private bus operators, and are identified by route numbers beginning with "N". All valid CityRail tickets for a destination (apart from single tickets) are accepted on NightRide services.[11] Bus stops and railway stations do not always perfectly coincide, but there is a reasonable approximation on most routes.

Network overview

The CityRail network is a hybrid of three different types of passenger railway: metro-style underground; suburban commuter rail and interurban.

For example, a person who lives in Blacktown, 30 km (19 mi) west of Sydney and works in the city centre 2 km (1 mi) from Sydney's Central Station can catch a CityRail suburban service from his/her local station. However, the train does not terminate at Central Station, instead proceeding onward into the underground network in Sydney's CBD and some inner city neighbourhoods without the need to change trains or buy tickets from a different railway organisation.

An R set leaves Artarmon railway station, on the North Shore Line

There is evidence this hybrid arrangement was deliberate. The design of the early electric carriages was developed as a combination of the high-capacity, low-boarding time of the New York Subways trains and the existing English long carriage design that was established in Australia's long-haul steam train system.[12] Those design principles have carried over to successive rolling stock.

CityRail also operates several interurban services that terminate at Central Station (though some services operate in the metro-style portions of the system in the peak hours). These lines stretch over 160 km (99 mi) from Sydney, as far north as Newcastle, as far west as Lithgow, as far south-west as Goulburn and as far south as Kiama and Port Kembla. Southern Highlands trains require a connection at Campbelltown as they run into the city during peak hours only.

An H Set waits to depart Epping Station

Regional services operate from the terminus station at Newcastle, with local electric services to the Central Coast and diesel services to Maitland. After Maitland, the DMUs travel either to Scone or Dungog, but most terminate at Maitland or Telarah. Another regional service operates as part of the South Coast Line, with DMUs between Kiama and Bomaderry-Nowra.

The hub of the CityRail system is Central Station, where most lines start and end. Trains coming from the Airport and East Hills Line and Bankstown Line, after travelling anticlockwise on the City Circle sometimes terminate upon arrival at Central and proceed to the Macdonaldtown Turnback. However, most trains continue on and become respective outward bound Inner West trains and South Line trains. The reverse applies for trains coming from the Inner West and South Lines, which, if not terminating, become outward bound trains on the Airport and East Hills line and Bankstown Line respectively. In the same manner, all trains from the Western Line or Northern Line become North Shore line trains once they reach Central and vice-versa.

As well as the intercity services mentioned above, local services also run in the Newcastle local area during off-peak times, as part of the Newcastle & Central Coast Line. Local services also run on the South Coast Line in the Wollongong local area, usually between Thirroul and Port Kembla.

Passenger Information Systems

The majority of CityRail stations are well equipped with electronic passenger destination indicator boards. These provide information on the current time, next three available services, time due to arrival, destination route and the number of train carriages available. Systems at Central station also produce a visual alert to passengers of train doors about to close during peak hour.

Due to the many differing types of stations that CityRail serves, their screens vary in form. For example, the vertical ceiling mounted screens are designed for higher patronage stations and appear on most city stations (mainly around the City Circle, the entire Eastern Suburbs railway line, the entire Epping to Chatswood railway line and Strathfield, whereas the dual horizontal screens appear in most other stations across the Sydney railway network.. CBSM (Custom Built Sheet Metal) was responsible for the manufacture of many indicator board encasings.[13]

Dual horizontal, ceiling mounted screens, standard type. Used at Airport stations as well. Photographed at Glenfield, an outer city suburban station.

Vertical painted white metal standing screens, station concourse/entrance type. Photographed at Glenfield, an outer city suburban station.

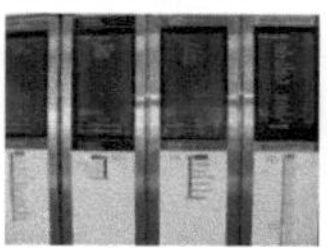

Vertical metallic standing screens, station concourse/entrance type. Photographed at Epping, an inner city suburban station.

Single, vertical ceiling mounted screens, larger passenger capacity station type. Photographed at Town Hall, an underground station located in the Sydney City CBD.

"Next Trains to Central" indicators. Photographed at Wynyard, an underground station located in the Sydney City CBD.

"Next trains to City" indicators. Photographed at Auburn, an inner city suburban station.

Intercity Indicator, includes Interstate CountryLink services. Photographed at Central

History

CityRail's origins go as far back as 1855 when the first public railway in New South Wales opened between Sydney and Granville, now a suburb of Sydney but then a major agricultural centre. The railway formed the basis of the New South Wales railways and was owned by the government. Passenger and freight services were operated from the beginning. The State's railway system quickly expanded from the outset with lines radiating from Sydney and Newcastle into the interior of New South Wales, with frequent passenger railway services in the suburban areas of Sydney and Newcastle along with less frequent passenger trains into the rural areas and interstate. All services were powered by steam locomotives, though in the 1920s petrol railcars were introduced for minor branch lines with low passenger numbers, both in metropolitan Sydney and rural areas.

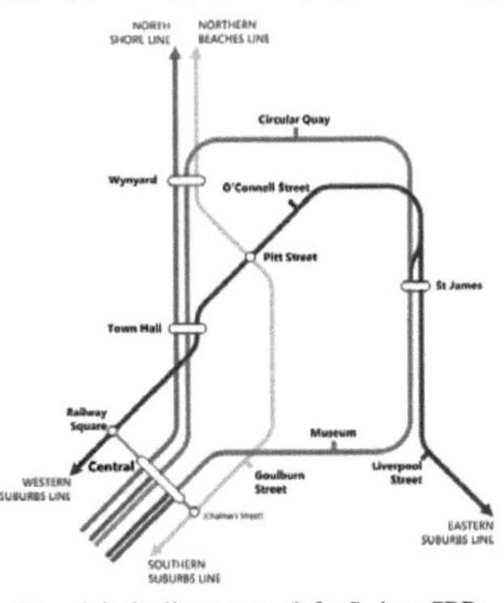

The original railway network for Sydney CBD planned by John Bradfield

The CityRail system as it exists today is to some extent the result of the vision and foresight of John Bradfield, one of Australia's most respected and famous civil engineers. He was involved in the design and construction of Sydney's underground railways in the 1920s and 1930s, but he is more famous for the associated design and construction of Sydney's greatest icon, the Sydney Harbour Bridge.

Electrification

New South Wales uses an overhead electrification system at 1 500 volts direct current. Whilst inferior to and more expensive than modern single phase alternating current equipment, it was in vogue during the 1920s and is generally sufficient for the operation of electric multiple unit trains. However, the introduction of powerful electric locomotives in the 1950s, followed by the Millennium Train in 2002, revealed drawbacks in this antiquated system of electrification. As the voltage is relatively low, high currents are required to supply a given amount of power, which necessitates the use of very heavy duty cabling and substation equipment (among the heaviest in the world). Until the retirement of electric locomotives from freight service, it was often necessary to observe a "power margin" to ensure that substations were not overloaded. This situation was similar to that which applied to The Milwaukee Road's 3 000 VDC electrification. Plans to electrify the Hunter Valley at 25 kV alternating current were abandoned in the 1990s. With private freight operation favouring diesel haulage, it is unlikely that the electrification will extend beyond its present outer-metropolitan limits in the foreseeable future.

Electrification came to Sydney's suburbs in 1926 with the first suburban electric service running between Sydney's Central Station and the suburb of Oatley approximately 20 km (12 mi) south of Sydney. In the same year, the first underground railway was constructed from Central Station to St James in Sydney's CBD . Electric trains that had previously terminated at the Central Station continued north, diving underground at the Goulburn Street tunnel portal, stopping at Museum underground station and then terminating at St James. Other lines were soon electrified. Also, in conjunction with the construction of the Sydney Harbour Bridge which opened in 1932, an additional underground line in downtown Sydney was constructed, connecting the North Shore line with Central Station via two downtown stations, Town Hall and Wynyard.

World War II interrupted programmes for further electrification, but an extensive electric network was in place in 1948.

Structure

The Public Transport Commission of New South Wales was created in 1972 by the merger of the Department of Railways, New South Wales and the New South Wales Department of Government Transport, which operated buses and ferries. It was broken up in 1980 into the State Rail Authority and Urban Transit Authority. CityRail was established in 1990 under the *Transport Administration Act (NSW) 1988* as a business unit of State Rail.

Challenges

The quality of the rail system is a matter of considerable political sensitivity. The performance of StateRail and RailCorp have been questioned in regards to safety, training, a politically-motivated focus on punctuality, management and workplace culture, with strong criticism from Justice Peter McInerny in his inquiries into the accidents at Glenbrook and Waterfall.[14] [15] Transport is the third largest area of public expenditure in NSW, after health and education. A newspaper distributed to commuters, *mX*, and the *Sydney Morning Herald's* "campaign for Sydney" kept transport at the top of the agenda ahead of the 2007 state election. In his 2003 interim report to the NSW Government, Tom Parry was highly critical of CityRail. "It is hard to believe that taxpayers or the state are getting the best possible value from the large amounts of money being spent each year," he wrote.[16]

Safety

The safety of the CityRail network was called into question by two fatal accidents. The second Glenbrook train disaster in 1999 killed seven people. In 2003, the Waterfall train disaster killed six. Inquiries were conducted into both accidents. Official findings into the latter accident also blamed an "underdeveloped safety culture." There has been criticism of the way CityRail managed safety issues that arose, resulting in what the NSW Ministry of Transport called "a reactive approach to risk management." CityRail has launched public information campaigns regarding railway trespassing, prams and strollers, and falling between the platform and the train.[17]

A typical message asking passengers to stand behind the yellow line, and tactile paving near the edge of the platform

Crime and terrorism

Crime committed on railway property has decreased by 32.9% since 2002, which RailCorp attributes to the deployment of some 600 Transit Officers across the network.[18] Most stations now have emergency "help points" to put passengers in immediate contact with authorities should an incident occur. All stations are covered by closed-circuit television surveillance. However a large amount of graffiti is still evident on some trains and the depots.

In recent years, concerns over terrorism have played a role in the management of the network. CityRail and other public transport providers participate in an ongoing public terrorism awareness campaign, *If you see something, say something*, adapted from a similar campaign in New York.[19]

An emergency help point

Integrated ticketing system

Tcard was an integrated inter-modal stored-value ticketing project, similar to London's Oyster Card, originally intended to be in place before the 2000 Sydney Olympics. The project experienced many delays and was the source of acrimony and lawsuits between the developer ERG and the state government. In 2007 the state government terminated the project and the $64m invested in it has been written off. ERG and the independent regulator attributed some of the delays to CityRail's complex fare structure.[20]

In May 2010 it was announced that a contract has been signed for the design and implementation of a new Electronic Ticketing System.[21]

Overloading

In 2008 overloading of trains was found by the Independent Pricing and Regulatory Tribunal[22] to be a significant cause of delays. Ten of the 13 lines have peak loads over 135% of capacity, and three in ten trains during the morning peak are officially overcrowded.[23]

Public perception

One result of CityRail's increasing problems has been a sharp rise in public complaints and attacks against staff,[24] with a Boston Consulting Group report claiming staff were actively hiding from irate customers wishing to complain about the service. The highly negative public perception of transit officers acting as ticket inspection officers and charging significant on-the-spot fines has also led to the organisation introducing anti-spitting fines and signage requesting commuters not abuse staff.[25]

Future development

The CityRail network is undergoing a process of expansion in response to concerns that rail services are inadequate in Western Sydney. At present, the Transport Construction Authority is undertaking or planning several construction projects for CityRail. Currently under construction is the South West Rail Link which will extend the network to Leppington.

The Government of New South Wales announced in 2003 that it intended to separate the existing CityRail lines into five independent lines with more reliable and frequent services. The project is called "Rail Clearways", and the five new sectors are listed as the Illawarra and Eastern Suburbs Line, the Bankstown Line, the Campbelltown Express Line, the Airport & South Line and the North-West Lines.

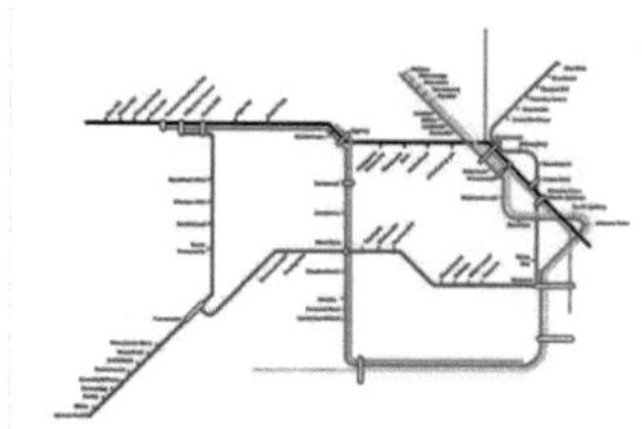

Partial diagram of possible 2050 network under the proposed Christie report

Related to the Clearways Project is the construction of the 13 km (8 mi) Epping to Chatswood Line which runs between Epping on the Northern Line and Chatswood on the North Shore line, with three new stations in between, servicing the growing industrial and commercial area of North Ryde, Macquarie University, and Macquarie Shopping Centre. The line opened on 23 February 2009.[26]

There are plans to install ETCS across much of the network, starting with the Berowa-Wyong route in 2013.[27]

Changes to CityRail

- **May 2000**: New timetable, Airport line opens.
- **July 2002**: Millennium trains are entered into service operating on the Airport & East Hills, South, Bankstown, & Inner West Lines. This is the first CityRail train to have new features: CCTV security cameras, Automatic DVAs, Wheelchair Ramps and indicator boards.
- **September 2005**: New timetable for all lines.
- **May 2006**: New timetable for Eastern Suburbs, Illawarra and South Coast Line, including revised changes on all lines.
- **December 2006**: Oscars are entered into service starting with the South Coast Line.
- **September 2007**: DVAs are introduced on all trains waiting for the doors to close with the catch-phrase "Doors closing, please stand clear".
- **October 2007**: Hunter Railcars are entered into service.
- **December 2007**: Oscars expanded its service on the Central Coast Line via North Shore Line.
- **February 2009**: Epping to Chatswood line opens with shuttle services.
- **October 2009**: New timetable, Epping to Chatswood line integrated to the Northern Line. CityRail launches the Revesby, Chatswood and Epping terminus, Oscars expanded its off-peak service with K-Sets on the Northern Line as a replacement to Tangaras which are abandoned from the Epping to Chatswood Line due to steepness and also the R & S Sets because of noisiness of the tunnel.
- **October 2010**: New timetable for Eastern Suburbs, Illawarra and South Coast Line (Which includes extra services to Cronulla, extra weekend services for the Eastern Suburbs and Illawarra, & South lines as well as more direct fast weekend services between Kiama and Bondi Junction on the South Coast line) including revised changes on all lines.
- **July 2011**: The first of the Waratah trains entered service.
- **2011-2014**: Waratah trains will be entered into service.

See also

- List of Sydney railway stations
- Metro Light Rail
- Metro Monorail
- Railways in Sydney
- Southern Sydney Freight Line
- Sydney underground railways

References

[1] http://www.cityrail.info/

[2] Powerhouse Museum - First train between Sydney - Parramatta (http://www.powerhousemuseum.com/exhibitions/locomotive1.php), accessdate =30 December 2010

[3] http://www.cityrail.info/about/facts

[4] "Train Fleet Maintenance" (http://web.archive.org/web/20080313154016/http://www.cityrail.info/aboutus/our_performance/train_maintenance.jsp). Cityrail. 2006-06-01. Archived from the original (http://www.cityrail.info/aboutus/our_performance/train_maintenance.jsp) on 2008-03-13. . Retrieved 2008-05-18.

[5] Department of Railways, New South Wales: Working of Electric Trains, 1965

[6] Parry, Thomas G. (2003-12-01). "Ministerial Inquiry into Sustainable Transport in New South Wales" (http://www.transport.nsw.gov.au/inquiries/parry-final-report.html). New South Wales Ministry of Transport. . Retrieved 2008-05-18.

[7] "CityRail - Tickets & Fares" (http://www.cityrail.info/fares/calculator.jsp?from=344&to=19). RailCorp. 2009. . Retrieved 2009-06-09.

[8] "Our Performance - On-Time Running and Service Reliability" (http://web.archive.org/web/20080218030727/http://www.cityrail.info/aboutus/our_performance/summary_otr.jsp). Cityrail. Archived from the original (http://www.cityrail.info/aboutus/our_performance/summary_otr.jsp) on 2008-02-18. . Retrieved 2008-05-18.

[9] "Aussie train services 'among world's worst'" (http://www.news.com.au/story/0,23599,21418282-2,00.html). News.com.au. March 21, 2007. . Retrieved 2008-01-11.

[10] http://www.cityrail.info/stations/network_map

[11] "Sydney's Waratah train" (http://www.cityrail.info/nightride/index.jsp). CityRail. . Retrieved 29 January 2011.

[12] "Historic Electric Traction "What Is So Special About Sydney Single Deckers?", paragraph 2" (http://www.het.org.au/blurp.htm). .

[13] http://www.cbsm.com.au/clients.html

[14] Special Commission of Inquiry into the Waterfall Rail Accident, Final Report, Volume I, January 2005, The Honourable Peter Aloysius McInerny QC

[15] Railway Safety: Interlocking and Train Protection, Ian Macfarlane, 2004

[16] "Ministerial Inquiry into Sustainable Public Transport" (http://www.transport.nsw.gov.au/inquiries/parry-final-report.pdf). New South Wales Government & Tom Parry. 9 December 2003. . Retrieved 30 March 2008.

[17] "Training Rules" (http://web.archive.org/web/20080419113045/http://www.cityrail.info/training_rules/training_rules.jsp). Cityrail. Archived from the original (http://www.cityrail.info/training_rules/training_rules.jsp) on 2008-04-19. . Retrieved 2008-05-18.

[18] "RailCorp Annual Report 2006-2007" (http://www.railcorp.info/__data/assets/pdf_file/0003/5493/RailCorp_Annual_Report_2006-2007.pdf) (PDF). RailCorp. 2007-10-31. . Retrieved 2008-03-04.

[19] "CityRail: Security: If you see something, say something" (http://web.archive.org/web/20070828220439/http://www.cityrail.info/security/see_something.jsp). Rail Corporation New South Wales. Archived from the original (http://www.cityrail.info/security/see_something.jsp) on 28 August 2007. . Retrieved 14 September 2007.

[20] "$64m Tcard fiasco over" (http://www.smh.com.au/news/national/64m-tcard-fiasco-over/2007/11/09/1194329499286.html). Sydney Morning Herald. 2007-11-09. . Retrieved 1 August 2008.

[21] "Electronic ticketing contract signed" (http://www.nsw.gov.au/news/electronic-ticketing-contract-signed). NSW Government. .

[22] http://www.ipart.nsw.gov.au/welcome.asp

[23] "Stuffed: Cityrail's timetable woe" (http://www.smh.com.au/news/national/cityrails-timetable-woe/2008/06/06/1212259115348.html). Sydney Morning Herald. 7 June 2008. . Retrieved 1 August 2008.

[24] "CityRail complaints" (http://www.smh.com.au/news/national/cityrail-complaints/2008/10/09/1223145541968.html). The Sydney Morning Herald. 2008-10-10. .

[25] http://www.cityrail.info/travelling_with/conditions_of_travel/fines

[26] "Epping to Chatswood Rail Link - CityRail website" (http://www.cityrail.info/news/projects/ecrl/for_customers). .

[27] "Railway Gazette: NSW awards first ETCS contract" (http://www.railwaygazette.com/nc/news/single-view/view/nsw-awards-first-etcs-contract.html). . Retrieved 2011-02-13.

Other references

- RailCorp Annual Report 2005-2006
- Interim Report of the Ministerial Inquiry into Sustainable Transport in New South Wales
- Details and map of Clearways (http://www.cityrail.info/news/clearways.jsp)
- North West Rail Link (http://www.railcorp.info/about_railcorp/major_projects/north_west_rail_link)
- South West Rail Link (http://www.planning.nsw.gov.au/plansforaction/pdf/swrl.pdf)
- Christie Report (http://www.aptnsw.org.au/christie/) - Long Term Strategic Plan for Rail.
- NSW Audit Office: Managing Disruption to CityRail services. (http://www.audit.nsw.gov.au/publications/reports/performance/2005/cityrail/cityrail-contents.html)

External links

Official

- CityRail (http://www.cityrail.info/) - CityRail Website
- CityRail's Fleet (http://www.cityrail.info/about/fleet/) - CityRail webpage of its rolling stock.
- RailCorp Homepage (http://www.railcorp.info/)
- Rail Infrastructure Corporation (http://www.ric.nsw.gov.au/)
- State Transit Authority (http://www.sta.nsw.gov.au/)

Enthusiast

- Railpage Australia (http://www.railpage.com.au/) - General news and discussion about Australian railways.
- Sydney Electric Train Society (SETS) Incorporated (http://www.sets.org.au/) - Established in 1991 for the preservation and operation of Sydney's vintage single deck electric trains and electric locomotives 4615 and 8606.
- Trendy's Trainpage (http://web.archive.org/web/20091027063054/http://geocities.com/trendy_rechauffe/) - Facts and photos of the trains and railways of Sydney, Australia
- The Intercity Platform (http://www.theintercityplatform.com/) - Images (and videos) of the CityRail Fleet.
- Historical NSW Rail Timetables (http://web.archive.org/web/20091027045445/http://geocities.com/nswrail/)- Historical Sydney Railway Timetables

South Coast railway line, New South Wales

South Coast line	
Mode	Interurban rail line Coach service
Owner	CityRail
Operator(s)	CityRail
Connects	Central Wollongong Port Kembla (branch) Dapto Bundanoon (by coach) Kiama Bomaderry (Nowra)
Length	153 km
Stations	40
Fleet	H, L, G, V sets & Endeavour railcars
Depot(s)	Mortdale for L and G sets, Flemington for V sets, Eveleigh for H sets
Line colour	Blue/Grey
Key dates	
1887	Opened

The **South Coast Line** is in the intercity region of Sydney's CityRail services. It serves the coastal region to the south of the Sydney metropolitan area including the Illawarra region, most notably the regional city of Wollongong, and extended services reach as far as Nowra in Shoalhaven.

Line naming

The line is operationally and historically known as the *Illawarra Line* throughout its length from the Illawarra Junction at Redfern to its terminus in Bomaderry. CityRail currently markets the suburban services to Waterfall and Cronulla as the *Illawarra line* and interurban services south to Wollongong and Bomaderry as the *South Coast line*.

Description of route

Suburban section

[1] The **South Coast (Illawarra) line** commences at the *Illawarra Junction* at Redfern. Here, a dive-under allows inter-city services from the South Coast line to cross underneath the main suburban railway lines to access Sydney Terminal. From the Illawarra junction, four tracks head south through the stations of Erskineville and St Peters to Sydenham. Immediately north of Erskineville station, the Illawarra lines are connected to the *Illawarra Relief Lines* which emerge from underground. These lines form the Eastern Suburbs line which opened in 1979. Heading south from Erskineville, the eastern-most pair of tracks are the *Up and Down Illawarra* lines which usually carry the Illawarra line passenger services.

The western-most pair of tracks are the *Up and Down Illawarra local tracks* which usually carry Bankstown and East Hills line express trains. To the west of the four tracks between Erskineville and Sydenham lies a reservation for a further pair of tracks with partially constructed platforms at Erskineville and St Peters stations. There are current plans to complete these tracks under the Rail Clearways plan, these tracks will be known as the *Up and Down Illawarra relief lines*.[2]

At Sydenham, six platforms are provided, with Bankstown line services generally using the western-most pair (platforms 1 and 2), East Hills peak hour services using the inner pair (platforms 3 and 4) and Illawarra line services using the easternmost pair of platforms (platforms 5 and 6). South of Sydenham, the Bankstown line branches off in a westwards direction. The Botany Goods Line crosses over the Illawarra line via a flyover. The line then reaches Tempe station, before crossing the Cooks River.

South of the Cooks River lies Wolli Creek station, where the East Hills line branches off to the west. The Illawarra line continues south as four tracks through a rock cutting to the stations of Arncliffe (2 island platforms), Banksia (2 side platforms and an island platform) and Rockdale. Rockdale station has five platforms, platform 1 (the most westerly platform) is currently unelectrified and disused but was previously a terminating point for electric passenger trains. South of Rockdale, the line passes through Kogarah station (one island platform 2 & 3 and 2 side platforms 1 & 4) which has a shopping centre built overhead. The line the makes a westerly turn, heading through Carlton and Allawah stations (both with two island platforms).

The next station is Hurstville (2 island platforms), which is where the four track section ends and terminating facilities are provided. Like Kogarah, Hurstville has a shopping centre built above the platforms. South of Hurstville, the line becomes 2 tracks with bidirectional signalling. The line passes through Penshurst and Mortdale (island platforms). At Mortdale is the Mortdale maintenance depot which lies on the eastern side of the tracks with access points from the south of the station. The line then continues to Oatley which has an island platform and a set of points allowing trains to be turned-back. The line then crosses the Georges River over the Como Bridge, which opened in 1972

replacing an older single track bridge which still exists to the east of the present structure and is used as a cycleway. The line enters the Sutherland Shire, passing through Como station (which was moved to its present, new site with the opening of the new bridge in 1972), and Jannali (side platforms) before reaching Sutherland.

At Sutherland, three platforms are provided. The Cronulla Line branches off in an eastwards direction south of the station. The former short branch line to Woronora Cemetery branched in a westerly

Como Bridge

direction at the south of the platforms. The line opened on 28 July 1900 and closed on 27 August 1944.[3] The line then continues south through Loftus, Engadine, and Heathcote (all side platform stations). South of Loftus, the former Royal National Park line branched off, this has now been converted into a tram line connecting to the Sydney Tramway Museum, and connections to the mainline have been severed. The final station for the operation of suburban services is Waterfall station, which is an island platform. At Waterfall, there is a train stabling yard and a train turnback (shunting road) south of the station. South of Waterfall is the site of the 2003 Waterfall train disaster.

Inter-urban section

The line then heads south through the challenging terrain of the Royal National Park and Illawarra escarpment. The line makes a steep descent down to Wollongong. The original alignment through the towns of Helensburgh and Lilyvale which opened in 1888 was bypassed by a new route in 1915. A new station at Helensburgh (island platform) was subsequently opened with the new alignment.[4] A set of points allows the turnback of trains at Helensburgh. The line then proceeds through several tunnels down the Illawarra escarpment through the hamlets of Stanwell Park and Coalcliff.

South of Coalcliff, the line becomes single track as it passes through the Clifton Tunnel, before becoming double track again near Scarborough station. The line then proceeds south through the northern suburbs of Wollongong, then Wollongong and its southern suburbs. A terminating platform is provided at Thirroul, which is used to terminate peak hour services from Sydney, as well as local services.

At Coniston south of Wollongong, an electrified branch line heads east to Port Kembla. At Unanderra, the line to Moss Vale branches off to head west over the Illawarra escarpment to join the Main South line. Double tracking ends at Unanderra, and the line continues south as a single line track. The line continues south through Kembla Grange where a simple platform serves the Kembla Grange racecourse. The line then reaches Dapto where a passing loop is provided. Dapto was the southern extent of electrification until 2001[5]

Coledale station

OSCar train at Scarborough

The line then passes south through the hamlet of Albion Park Rail (where another crossing loop is provided) to reach Kiama, the current extent of electrification. South of Kiama, the line continues as a single track non-electrified line through rolling dairy pastures via several tunnels to the towns of Gerringong and Berry before arriving at its terminus at the town of Bomaderry on the northern bank of the Shoalhaven River. At Bomaderry, sidings connect to The Manildra Group's starch mill.[6]

Port Kembla branch

At Coniston, an electrified branch line proceeds east to the industrial suburb of Port Kembla with three intermediate stations.

The line is double track as far as just west of Port Kembla North and is used by freight trains as well as local CityRail passenger services. A stabling yard is provided at Port Kembla for overnight storage of electric trains.

History

L set train at Cringila

A Sydney to Kiama railway was authorised by the New South Wales Parliament in April 1881.[7] Construction of the various sections was awarded by tender and commenced in October 1882.[7]

The Illawarra Line branched off the Main Suburban Railway south of Redfern, at Eveleigh (*Illawarra Junction*). The line opened to Hurstville in 1884, Sutherland in 1885, Waterfall in 1886 and Clifton through to Wollongong and

North Kiama (Bombo) in 1887.[8] The missing Waterfall to Clifton section comprised four large brick-arch culverts (and many small ones) and eight tunnels with a total length of over 4 km, delaying its opening until 1888.[7] Kiama and Bomaderry (servicing Nowra) opened in 1893.[8] The line was originally double track to Hurstville and continued as a single track, but was duplicated to Waterfall (except for the Como to Sutherland section) in 1890.[7] In 1886 a branch line was opened to Audley in the Royal National Park, which closed in 1991.

A steep ruling grade of 1 in 40 faced up (Sydney bound) trains almost all the way between Stanwell Park station and Otford. The main problem was the 1550 m long Otford Tunnel, which took the railway through Bald Hill from the coast at Stanwell Park to the Hacking River valley. The steep 1 in 40 grade and tight clearances meant that soot, smoke and heat could become unbearable, especially when a south-easterly wind blew into the southern portal or when a train stalled in the tunnel.[7] A Mr B Chamberlain wrote about a stalled passenger train in 1890:[7]

> Even with the windows closed, the carriages were filled with smoke and steam, women fainted and children screamed until the train backed down to Stanwell Park, and was finally staged up to Otford in two trips.

Regarding the crew, Chamberlain wrote:[7]

> While the passenger with closed windows in an up train had an unpleasant journey... the unfortunate enginemen underwent a shocking ordeal. On tender engines both knelt on the footplate, coats over heads, to breathe the air coming from under the engine, the apron plate being raised for this purpose. Though the air was hot from passing around or through the ash pan, it was none the less welcome.

Attempts were made to overcome the problem with a ventilation shaft and chimney in the early 1890s and a blower system installed in 1909.[7]

Many goods trains were routinely divided at Stanwell Park and taken through to Waterfall in stages, effectively increasing the number of train movements on the line. The increasing congestion and steepness led to construction of a double track deviation, which opened between Waterfall and Helensburgh in 1914, Helensburgh and Otford in 1915, and Otford and Coalcliff (bypassing the by now infamous Otford Tunnel) in 1920. The deviation avoided the steep grades with a more winding route featuring sharp curves, deep cuttings, new tunnels and a curved viaduct over Stanwell Creek that required 3 million bricks in its construction. Although the new route was 5 km longer it reduced the ruling grade from 1 in 40 to around 1 in 80.[7] Of the original eight tunnels in this section only the Clifton Tunnel remains in use and is the only single track section between Sydney and Unanderra.

In 1924, work began on a 57 km line connecting Unanderra with Moss Vale on the Great Southern line, which opened in 1932. It enabled the transportation of limestone from the Southern Highlands to the coast at Port Kembla. *See: Unanderra - Moss Vale railway line.*

Electrification of the South Coast Line was completed in stages reaching from Sydney to Loftus in 1926, Waterfall in 1980, Helensburgh in 1984, Wollongong in 1985-86, Dapto in 1993 and Kiama in 2001. The Kiama to Nowra section remains unelectrified.

Major structural problems with the Stanwell Creek viaduct were identified in late 1985, with one span close to collapsing and another badly cracked, requiring substantial repairs and stabilising work.[7]

On 31 January 2003, an Intercity Tangara passenger train (G7) derailed at high speed south of Waterfall station after its driver suffered a heart attack. The Waterfall train disaster resulted in seven fatalities and multiple injuries.

Passenger services

The South Coast line passenger services currently consist of electric double deck multiple unit trains that operate between Bondi Junction or Central and either Wollongong, Kiama or Port Kembla. Diesel shuttle trains connect at Wollongong or Kiama and operate to Nowra. Although electrified to Wollongong in 1985, several diesel trains operated between Sydney and Nowra until 1991, one of which was the *South Coast Daylight Express*, operated as a locomotive hauled train of Budd and Tulloch type passenger cars which included catering facilities.[9]

OSCAR train at Wollongong Station

References

[1] Illawarra Line. Railcorp track diagram, 09 September 2002

[2] "TIDC" (http://www.tidc.nsw.gov.au/ViewSite.aspx?PageID=471). *TIDC*. . Retrieved 14 December 2006.

[3] *The Woronora Cemetery Branch Railway* Neve, Peter Australian Railway Historical Society Bulletin, August, 1993 pp187-195

[4] *The Helensburgh Deviation* Singleton, C.C. Australian Railway Historical Society Bulletin May, 1966 pp97-106

[5] "RIC Annual report 2001-2002" (http://www.railcorp.info/__data/assets/file/753/RIC_Annual_Report_2001-2002.pdf) (PDF). *Railcorp*. . Retrieved 8 January 2007.

[6] "Submission of shoalhaven City, 9 May 2005" (http://www.aph.gov.au/house/committee/trs/networks/subs/sub044.pdf) (PDF). *Parliament of Australia*. . Retrieved 8 January 2007.

[7] Oakes, John (2009) [2003]. *Sydney's Forgotten Illawarra Railways* (2nd rev. ed.). Sydney: Australian Railway Historical Society, NSW Division. pp. 11, 12, 23, 24, 26, 54–56, 60, 73, 79–85. ISBN 978-0-9805106-6-9.

[8] "South Coast line" (http://www.nswrail.net/lines/show.php?name=NSW:south_coast). www.nswrail.net. . Retrieved 2006-11-26.

[9] Walters, C. *The Last Daylight*. Railway Digest 1991, ARHS NSW.

Albion Park Rail, New South Wales

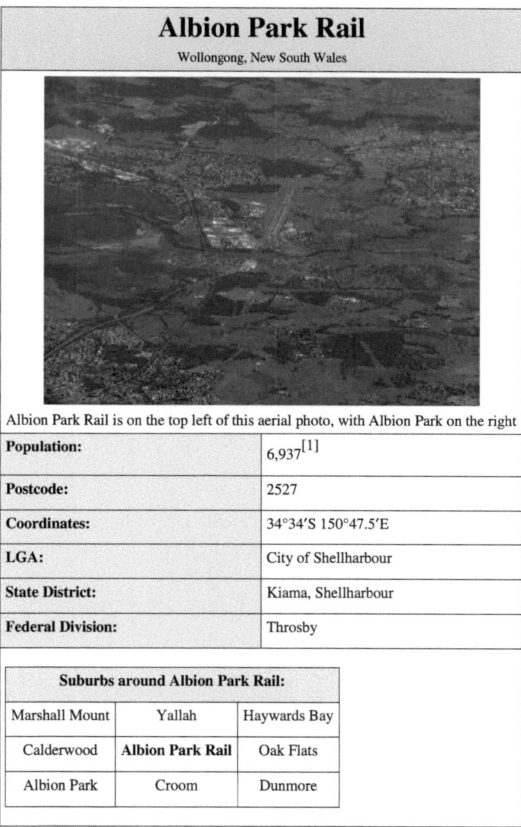

Albion Park Rail

Wollongong, New South Wales

Albion Park Rail is on the top left of this aerial photo, with Albion Park on the right

Population:	6,937[1]
Postcode:	2527
Coordinates:	34°34′S 150°47.5′E
LGA:	City of Shellharbour
State District:	Kiama, Shellharbour
Federal Division:	Throsby

Suburbs around Albion Park Rail:		
Marshall Mount	Yallah	Haywards Bay
Calderwood	**Albion Park Rail**	Oak Flats
Albion Park	Croom	Dunmore

Albion Park Rail is a suburb of Wollongong, New South Wales, Australia in the Local Government Area of Shellharbour. The South Coast railway line was opened to the railway station and Bombo in 1887. At the time the nearest town was Albion Park, several kilometres away. Over time, houses were built around the railway station, and Albion Park Rail developed into a town in its own right.

Albion Park Rail has a community area set up with a community centre, playing fields and a recently built skatepark. It only has one school but it accommodates many students from Pre-School to Year 6. Albion Park Rail is located along the shores of Lake Illawarra, near Koona Bay. It is a fifteen minute drive on the freeway to Wollongong.

Albion Park Rail is the site of the Illawarra Regional Airport which houses many rare planes including the "Connie".

The Illawarra Light Railway Museum located just south of the airport is dedicated to the preservation and display of historic light railway locomotives and rolling stock. It holds regular open days featuring light and miniature train rides.

External links

- Illawarra Light Railway Museum Society Homepage [2]

References

[1] Australian Bureau of Statistics (25 October 2007). "Albion Park Rail (State Suburb)" (http://www.censusdata.abs.gov.au/
ABSNavigation/prenav/LocationSearch?collection=Census&period=2006&areacode=SSC15011&producttype=QuickStats&
breadcrumb=PL&action=401). *2006 Census QuickStats.* . Retrieved 2008-08-09.
[2] http://www.ilrms.com.au/

Albion Park, New South Wales

<table>
<tr><td colspan="2" align="center">Albion Park
Wollongong, New South Wales</td></tr>
<tr><td colspan="2" align="center">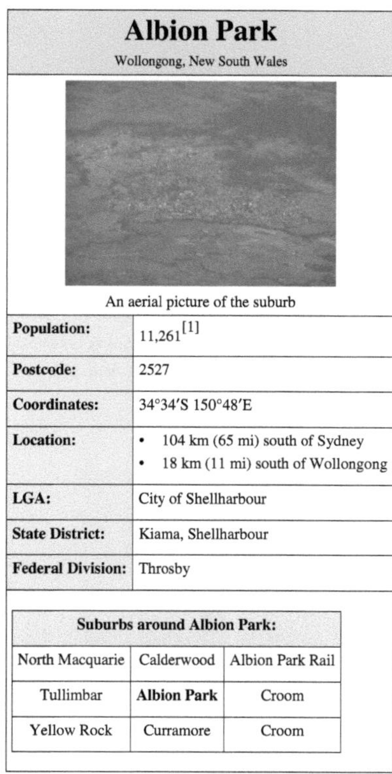
An aerial picture of the suburb</td></tr>
<tr><td>Population:</td><td>11,261[1]</td></tr>
<tr><td>Postcode:</td><td>2527</td></tr>
<tr><td>Coordinates:</td><td>34°34′S 150°48′E</td></tr>
<tr><td>Location:</td><td>• 104 km (65 mi) south of Sydney
• 18 km (11 mi) south of Wollongong</td></tr>
<tr><td>LGA:</td><td>City of Shellharbour</td></tr>
<tr><td>State District:</td><td>Kiama, Shellharbour</td></tr>
<tr><td>Federal Division:</td><td>Throsby</td></tr>
</table>

Suburbs around Albion Park:		
North Macquarie	Calderwood	Albion Park Rail
Tullimbar	**Albion Park**	Croom
Yellow Rock	Curramore	Croom

Albion Park is a southern suburb of Wollongong, New South Wales, Australia. Currently Albion Park has a population of approximately 11,261[1] and continues to expand by a rate of 12.5% per year. Albion Park is bordered by the Illawarra escarpment in the west, Dapto in the North, Jamberoo in the South and Shellharbour in the East.

History

The area around Albion Park was cleared by cedar cutters, and they were quickly followed by graziers who recognised the potential of this well-watered area. The arrival of the railway line in 1887, and the completion of the road through Macquarie Pass to the Southern Highlands hastened the town's growth. On the 12th of August 2009, it was voted the bogan capital of Australia.

Albion Park is the gateway in to the Southern Highlands, which the Southern Highlands is west of Albion Park.[2]

Etymology

The term "Albion", is an ancient name of Great Britain.[3] the name was given by the original land grant owner around 1815.

Infrastructure

Trains

In order to avoid the highest parts of the ridge between Albion Park and Minnamurra, the railway needed to keep closer to the shore of Lake Illawarra. After the opening of Albion Park railway station two km to the east in 1887, a new town grew up next to it, now known as Albion Park Rail.

Roads

The Illawarra Highway runs through Albion Park and joins the Princes Highway just north of the suburb. This road brings many people through the region. Albion Park is also the main link between Jamberoo Action Park and Sydney as the only way for people to get to Jamberoo coming from the north is to pass through Albion Park.

Education

There are currently 3 Primary Schools and 3 High Schools in Albion Park:

- Albion Park Public School
- Albion Park High School [4]
- Mount Terry Public School [5]
- St. Joseph's Catholic School, Albion Park [6]
- St. Paul's Catholic School, Albion Park [7]
- Illawarra Christian School [8] (Tongarra Campus)
- Tullimbar Public School
- Mount terry public school

Recreation

Albion Park is home to the Shellharbour Regional Sporting Complex, which consists of two all-weather hockey fields, several football (rugby league) fields, a multi-purpose indoor basketball stadium, a turf cricket field, an athletics field, a BMX track, a remote control car racing track, and several tennis courts. A cycleway also courses it's way through the Sporting Complex and links it to a polo field and a complex of soccer fields. Albion Park is also home to a Shellharbour Council Swimming Pool, which is open during the summer months and now costs a fee of $1 entry, to a pool paid for by the members of the town.

Clubs and Pubs

Albion Park is home to three clubs where people to eat and drink including:

- Albion Park Hotel
- Albion Park RSL [9]
- Albion Park Bowls and Recreation Club

Sport

Albion Park is also home to various sporting teams such as:

- Albion Park-Oak Flats Eagles
- Albion Park Hockey Club [10]
- Albion Park Bowls and Recreation Club
- Albion Park Cricket Club
- Albion Park White Eagles
- Albion Park Crows Junior AFL Club [11]

Museums

Albion Park has the Tongarra Museum, which showcases the history of the region, and the Historical Aircraft Restoration Society Museum located at the Illawarra Regional Airport.

Commercial area

Albion Park has a small shopping centre known as Centro Albion Park.

References

[1] Australian Bureau of Statistics (25 October 2007). "Albion Park (State Suburb)" (http://www.censusdata.abs.gov.au/ABSNavigation/ prenav/LocationSearch?collection=Census&period=2006&areacode=SSC15006&producttype=QuickStats&breadcrumb=PL& action=401). *2006 Census QuickStats.* . Retrieved 2008-11-20.

[2] VisitNSW – Albion Park (http://www.visitnsw.com/town/Albion_Park.aspx)

[3] Albion – Online Etymology Dictionary (http://www.etymonline.com/index.php?search=albion&searchmode=none)

[4] http://www.albionpk-h.schools.nsw.edu.au

[5] http://www.mtterry.nsw.edu.au

[6] http://www.stjosephs.woll.catholic.edu.au/

[7] http://www.stpaulsap.woll.catholic.edu.au/

[8] http://www.ics.nsw.edu.au/

[9] http://www.albionparkrsl.com.au/

[10] http://www.albionparkhockey.org.au/

[11] http://www.albionparkcrows.com.au/

Oak Flats railway station

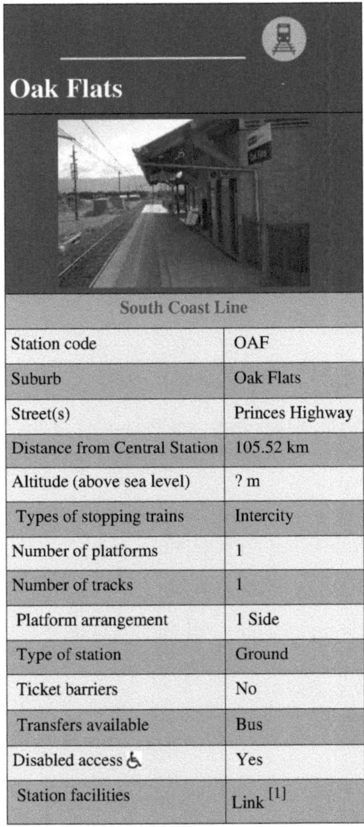

Station code	OAF
Suburb	Oak Flats
Street(s)	Princes Highway
Distance from Central Station	105.52 km
Altitude (above sea level)	? m
Types of stopping trains	Intercity
Number of platforms	1
Number of tracks	1
Platform arrangement	1 Side
Type of station	Ground
Ticket barriers	No
Transfers available	Bus
Disabled access ♿	Yes
Station facilities	Link [1]

Oak Flats is a station on the Cityrail South Coast line in New South Wales. It serves the suburb of Oak Flats. It is located on the single track section of the South Coast line between Unanderra and Bomaderry and has a single side platform.

History

A station opened on the site in 1925. In 2003 Oak Flats was rebuilt with a bus interchange about 400m south of its original site and the platform was moved to the eastern side of the track.[2]

Platforms and services

Platform	Line	Stopping Pattern	Notes and Comments
1	South Coast Line	Intercity services to Kiama, Dapto, Wollongong and Sydney Terminal. Peak Intercity services to Bomaderry.	

Transport links

Premier Illawarra runs nine routes to and from Oak Flats railway station:

- **37** - Wollongong Loop
- **43** - between Dapto and Port Kembla
- **51** - to University of Wollongong
- **52** - to Wollongong station
- **53** - to University of Wollongong
- **54** - to Wollongong station
- **57** - Wollongong Loop
- **70** - to Shellharbour district
- **76** - to Shellharbour district

Accessibility

Oak Flats has a side platform with near street level access. It therefore has close to Easy Access for wheelchairs.

Neighbouring stations

Preceding station	CityRail	Following station
Dunmore (Shellharbour) *towards Bomaderry (Nowra)*	South Coast Line	Albion Park *towards Central*

References

[1] http://www.cityrail.info/facilities/facilities.jsp?n=207&giveOutput=true&facility=

[2] "Oak Flats railway station" (http://www.nswrail.net/locations/photo.php?name=NSW:Oak+Flats:1&line=NSW:south+coast:0). www.nswrail.net. . Retrieved 2007-06-26.

Looking south

Dapto railway station

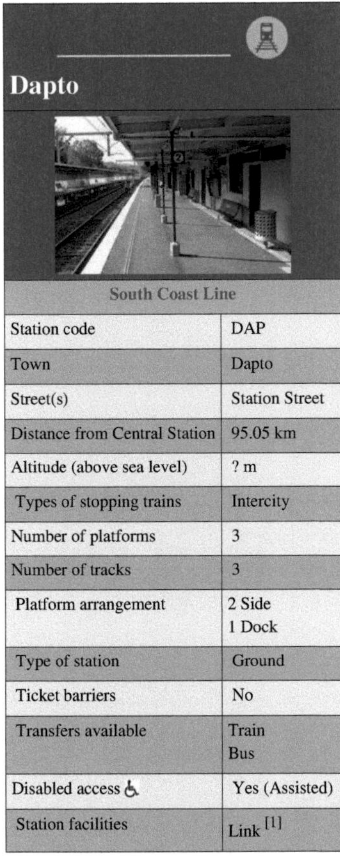

Station code	DAP
Town	Dapto
Street(s)	Station Street
Distance from Central Station	95.05 km
Altitude (above sea level)	? m
Types of stopping trains	Intercity
Number of platforms	3
Number of tracks	3
Platform arrangement	2 Side 1 Dock
Type of station	Ground
Ticket barriers	No
Transfers available	Train Bus
Disabled access	Yes (Assisted)
Station facilities	Link [1]

Dapto is a CityRail railway station on the South Coast railway line. It serves the suburb of Dapto the southernmost suburb of Wollongong. The original Dapto railway station was constructed in 1880. In November 2000, it was added to the state heritage list.[2]

The station has three platforms; two side platforms and a dock platform for terminating trains. Many trains from Sydney terminate at this station, as well as a limited number of electric and diesel services from Kiama and Bomaderry. Until 2001, it was the end point for electric trains on the line. Today the electric wires extend to Kiama.

Platforms and services

Platform	Line	Stopping Pattern	Notes and Comments
1	South Coast Line	Intercity services to Kiama. Peak Intercity services to Bomaderry.	
2	South Coast Line	Terminating services from Bomaderry; Intercity services to Wollongong and Sydney Terminal.	
3	South Coast Line	Terminating services; Intercity services to Wollongong and Sydney Terminal.	

Looking south

Transport links

Premier Illawarra runs two routes to and from Dapto railway station:

- **32** - to Wollongong.
- **37** - Wollongong Loop.

Other information

The Railway Line serves as a suburb border line of Dapto and Horsley. It is also the main entry point to Horsley and Dapto.

Neighbouring stations

Preceding station	CityRail	Following station
Albion Park *towards Bomaderry (Nowra)*	South Coast Line	Kembla Grange Racecourse *towards Central*

Station layout

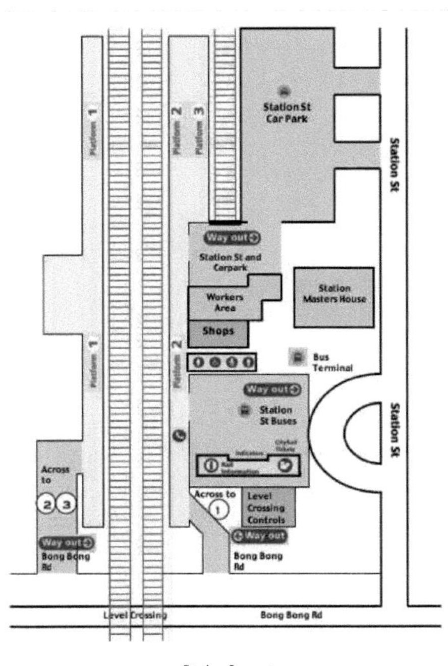

Station Layout

References

[1] http://www.cityrail.info/facilities/facilities.jsp?n=82& giveOutput=true&facility=

[2] "Dapto Railway Station group" (http://www.heritage.nsw. gov.au/07_subnav_02_2.cfm?itemid=5011984). State Heritage Register. . Retrieved 18 November 2010.

Bomaderry railway station

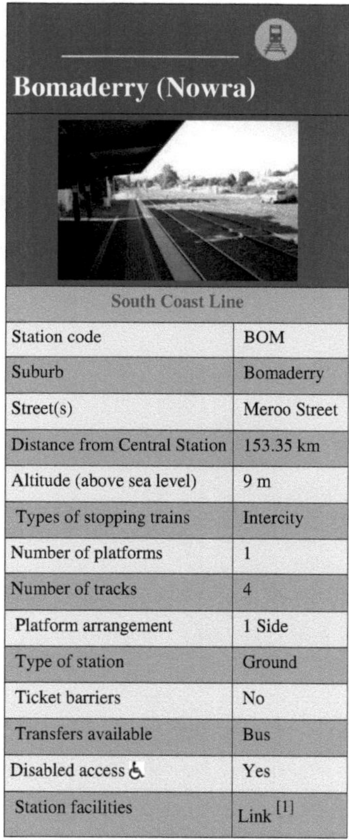

Station code	BOM
Suburb	Bomaderry
Street(s)	Meroo Street
Distance from Central Station	153.35 km
Altitude (above sea level)	9 m
Types of stopping trains	Intercity
Number of platforms	1
Number of tracks	4
Platform arrangement	1 Side
Type of station	Ground
Ticket barriers	No
Transfers available	Bus
Disabled access &	Yes
Station facilities	Link [1]

Bomaderry (Nowra) is a CityRail railway station on the South Coast railway line. The station opened on June 2, 1893.[2] It serves Bomaderry, which is a suburb of Nowra but on the northern side of the Shoalhaven River. The line was intended to be built across to Nowra proper, but the bridge was used for vehicle traffic instead.[3] No railway bridge crossing the Shoalhaven River at Nowra was ever built and to this day train services terminate on the northern side of the river. Bus services connect with the trains to the town centre.

The station has one side platform. There is a nearby paper mill siding for freight trains.

It was intended that railway be extended over old bridge (near side)

Derelict bridge over Bomaderry Creek, south of station

Station

Platform	Line	Stopping Pattern	Notes and Comments
①	South Coast Line	Intercity services to Kiama, Dapto and Wollongong.	Connecting Services at Kiama for Sydney Terminal.

Transport links

Nowra Coaches:

- **724** - to Shoalhaven University.
- **732** - to Basin View.
- **733** - to Wreck Bay.

Culburra Coaches:

- **729** - to Orient Point.

Kennedys Bus Service:

- **728** - to Greenwell Point.
- **809/10** - to Moss Vale.

Priors Scenic Express:

- **northbound**- to Bowral, Mittagong, Greater Sydney.

- **southbound**- to Ulladulla, Batemans Bay, Moruya, Tuross, Narooma.

Neighbouring stations

Preceding station	CityRail	Following station
Terminus	South Coast Line	Berry *towards Central*

References

[1] http://www.cityrail.info/facilities/facilities.jsp?n=35&giveOutput=true&facility=

[2] Bozier, Rolfe, *"New South Wales Railways: Bomaderry Railway Station"* (http://www.nswrail.net/locations/show. php?name=NSW:Bomaderry). Accessed 26 June 2007.

[3] " Nowra Bridge over the Shoalhaven River (http://www.rta.nsw.gov.au/cgi-bin/index.cgi?action=heritage.show&id=4301658)", *Roads and Traffic Authority*, 30 March 2004. Retrieved on 2008-11-20

Shellharbour, New South Wales

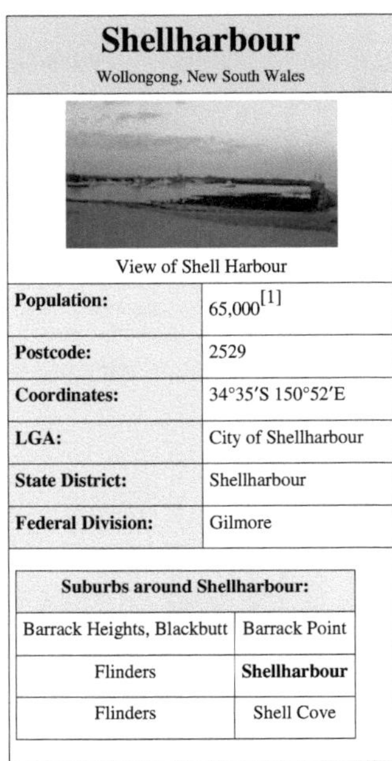

Shellharbour	
Wollongong, New South Wales	
View of Shell Harbour	
Population:	65,000[1]
Postcode:	2529
Coordinates:	34°35′S 150°52′E
LGA:	City of Shellharbour
State District:	Shellharbour
Federal Division:	Gilmore

Suburbs around Shellharbour:	
Barrack Heights, Blackbutt	Barrack Point
Flinders	**Shellharbour**
Flinders	Shell Cove

Shellharbour is a southern beachside suburb of Wollongong, located in the Illawarra region of New South Wales, Australia. It also gives its name to the Local Government Area, City of Shellharbour.

The suburb is centred on the small recreational harbour named Shell Harbour. It has two main beaches, Shellharbour Beach, which runs to Barrack Point and Shellharbour South Beach, which runs toward Bass Point.

Shellharbour is undergoing extensive development with many new retail outlets and units being constructed. The main street is Addison Street, with many sidewalk cafés and shops, running through the town and ending with the Ocean Beach Hotel opposite the harbour. Adjacent to the harbour is the Beverley Whitfield saltwater swimming pool and across from the Shellharbour Beach facilities is the Beverley Whitfield park, containing the Tom "Scout" Whilloughby cricket oval.

The area was inhabited by indigenous Australians for thousands of years. European habitation began from about 1817 onwards.[2] Shellharbour was originally known as *Yerrowah* and later as *Peterborough*.[3]

External links

- Shellharbour City Council website [4]
- Tourism Shellharbour website [5]
- Shellharbour Workers' Club website [6]
- Shellharbour City [7]

References

[1] Australian Bureau of Statistics (25 October 2007). "Shellharbour (State Suburb)" (http://www.censusdata.abs.gov.au/ABSNavigation/ prenav/LocationSearch?collection=Census&period=2006&areacode=SSC15236&producttype=QuickStats&breadcrumb=PL& action=401). *2006 Census QuickStats*. . Retrieved 2008-03-16.

[2] "Our History" (http://www.shellharbour.nsw.gov.au/default.aspx?WebPage=115). Shellharbour City Council. . Retrieved 2008-03-16.

[3] "Shellharbour" (http://www.gnb.nsw.gov.au/name_search/extract?id=TRqwXttLKW). Geographical Names Board of New South Wales. . Retrieved 2006-10-23.

[4] http://www.shellharbour.nsw.gov.au/

[5] http://www.tourismshellharbour.com.au/

[6] http://www.shellys.com.au/

[7] http://shellharbour.daveict.com/

Wollongong railway station

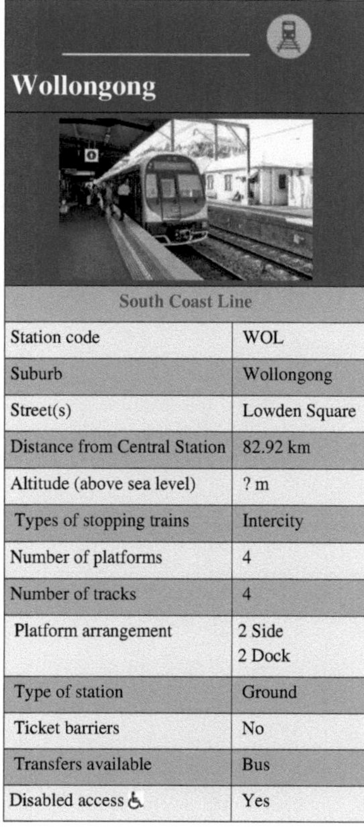

South Coast Line	
Station code	WOL
Suburb	Wollongong
Street(s)	Lowden Square
Distance from Central Station	82.92 km
Altitude (above sea level)	? m
Types of stopping trains	Intercity
Number of platforms	4
Number of tracks	4
Platform arrangement	2 Side 2 Dock
Type of station	Ground
Ticket barriers	No
Transfers available	Bus
Disabled access	Yes

Wollongong Station is a railway station on the South Coast Line of the CityRail outer suburban network, serving the central business district of the Illawarra region's major centre of Wollongong.

The station consists of two side platforms. It receives, on average, one off-peak service per hour travelling to/from Sydney on weekdays (with more services during peak hours) and one per hour on weekends. The station has two carparks and is situated just south of Jubilee Bridge on Crown Street. The Wollongong station is a CityRail 'Easy access' station, as it has a ramp and lift to all platforms, and has street access to ensure it is wheelchair accessible.

History

The original Wollongong railway station (Platform 2) was constructed in 1887. The Platform 2 refreshment room, which is still in use today, was constructed in 1890. The Platform 1 station and refreshment room were constructed in 1923 and 1926 respectively. These buildings, as well as the 1950's train crew building and 1928 Crown Street overbridge are listed on the NSW State Heritage Register.[2]

The station was given a major upgrade in 2005, adding new facilities such as an elevator on either platform.

On 14 August 2010, in addition to its original carpark, a new commuter car park which provides 365 free car spaces including disabled and motorcycle parking, as well as bicycle storage, was opened.[3] [4]

Platforms and service

Wollongong is the starting point of a branch service to Port Kembla, serving the Illawarra's main industrial district, although the line does not actually branch until the next station south, Coniston. The branch is served by roughly one off-peak train per hour between Wollongong and Port Kembla on weekdays, and an hourly service on weekends. A few additional services run during peak hours, including a couple of services running directly to and from Central, some continuing along the suburban Eastern Suburbs Line.

Platform	Line	Stopping Pattern	Notes and Comments
1	South Coast Line	Intercity services to Thirroul, Helensburgh, Waterfall and Sydney Terminal; terminating services.	Peak hour intercity services to Bondi Junction.
Dock platform	South Coast Line	Terminating services	
2	South Coast Line	Intercity services to Port Kembla, Dapto and Kiama; terminating services.	
Dock platform	South Coast Line	Limited intercity services to Bomaderry (Nowra)	

Transport links

Premier Illawarra:

- **37** - Lake Illawarra Loop
- **51** - between Wollongong University and Albion Park
- **52** - between Wollongong University and Oak Flats station
- **53** - between Wollongong University and Oak Flats station
- **54** - Between Wollongong University and Oak Flats station
- **57** - Lake Illawarra Loop
- **65** - between North Beach and Port Kembla station

Busways:

- **887** - to Campbelltown
- **887X** - to Campbelltown (Express via M1 Motorway)

Dock platform

Preceding station	CityRail	Following station
Coniston *towards Bomaderry (Nowra) or Port Kembla*	South Coast Line	North Wollongong *towards Central*

References

[1] http://www.cityrail.info/stations/station_details?stationId=235=

[2] "Wollongong Railway Station group" (http://www.heritage.nsw.gov.au/07_subnav_02_2.cfm?itemid=5012291). State Heritage Register. . Retrieved 18 November 2010.

[3] NSW Minister for Transport (October 14, 2010). "New Wollongong commuter car park opens" (http://www.transport.nsw.gov.au/sites/default/files/releases/101014_wollongong_commuter_car_parks_opens.pdf) (PDF). . Retrieved November 3, 2010.

[4] Matthew Jones (October 13, 2010). "New car park to assist rail commuters" (http://www.illawarramercury.com.au/news/local/news/general/new-car-park-to-assist-rail-commuters/1967672.aspx). . Retrieved November 3, 2010.

External links

- Cityrail station details for Wollongong station (http://www.cityrail.info/stations/station_details?stationId=235)

Dapto, New South Wales

<table>
<tr><td colspan="3" align="center">Dapto
Wollongong, New South Wales</td></tr>
</table>

Population:	10,480[1]
Postcode:	2530
Coordinates:	34°29.5′S 150°47.5′E
LGA:	City of Wollongong
State District:	Shellharbour
Federal Division:	Throsby

Suburbs around Dapto:		
Kembla Grange	Brownsville	Kanahooka
Horsley, Cleveland	**Dapto**	Koonawarra
Penrose	Yallah	Yallah

Dapto is a southern suburb of Wollongong in the Illawarra region of New South Wales, Australia, located on the western side of Lake Illawarra and covering an area 7.15 square kilometres in size. According to the 2006 ABS Census, the suburb currently has a total population of 10,480.[1]

History

The name Dapto is said to be an Aboriginal word either from *Dabpeto* meaning "water plenty", or from *tap-toe* which described the way an lame Aboriginal elder walked.[2] The suburb was officially founded in 1834, when George Brown transferred the Ship Inn from Wollongong to Mullet Creek Farm, in an area now named in his honour as Brownsville. After an unsuccessful attempt at wheat growing in the 1850s, Dapto embraced the dairy industry. In 1887 the railway opened and a butter factory was established. This began a transformation of Dapto and the town centre shifted south to where the new station was located. The Australian Smelting Company's works were established on Kanahooka Road and employed over 500 men. A railway, operated by the Illawarra Harbour and Land Corporation Limited, connected the smelter with the Government railway at Dapto.[3] By 1903 the Commissioner for Railways declared that Dapto was the most valuable station on the Illawarra line, its traffic being double that of Wollongong .

Further information

On Koonawarra Point is Mount Brown, a hill protected by a nature reserve for its important habitat. The eastern side and summit is owned by the now closed power station. Dapto used to be home to a smelting industry but this has now closed. Most of the town is residential with several commercial areas, the main one, including Dapto Mall, is on the Princes Highway near the station. At the station there is a mural depicting old and modern trains. Horsley is reached via a road from the station. A small area of the escarpment west of Dapto is protected as part of the Illawarra Escarpment State Recreation Area, though not accessible to the public. The escarpment west of Dapto includes Mount Bong Bong, the site of an aeroplane crash when the controllers mistook Lake Illawarra for Botany Bay. A plaque commemorates this event in Dapto. Much of the land to the west is still used as farmland, though other ventures such as the shooting grounds exist. The land to the west is noticeably more hilly than the plain to the north of Wollongong.

The South Coast line electric rail service terminated in Dapto prior to electrification being extended to Kiama in 2001. The area of Dapto west of this rail line, formerly pastoral land, has undergone significant expansion over the past 20 years. This area has become the fastest growing part of the greater Wollongong region and consists of a sprawling, medium-density suburbia known as Horsley, now officially a suburb in its own right.

Perhaps the area's most famous attraction is the Dapto Greyhound Racing Club [4], colloquially and affectionately known as the 'Dapto Dogs'. Race meetings occur from 6pm to 11pm every Thursday evening. On the 1st of November WIN News bulletin it was announced plans for a motel at the site were under way, as well as improvement of the track.

The Dapto rugby league team is known as 'The Canaries' and compete in the local Tooheys League. One of the most famous grand finals was in 1968 when Lionel Simmons kicked the winning field goal within seconds on the full time bell. Lionel Simmons is now a local legend.

Dapto was once, during an episode of the ABC TV show The Aunty Jack Show, part of a parody of the nearby city of Wollongong. The lyrics of the Lucky Starr song I've Been Everywhere were changed so that, instead of listing a wide range of Australian towns, the song said "I've been to Wollongong, Wollongong, Wollongong, Wollongong, Wollongong, Dapto, Wollongong, Wollongong (etc)".

Places of local interest

Dapto Mall

Dapto Mall is a shopping centre in the Dapto CBD. It has a vast range of stores including Coles Supermarket, Woolworths Store, Australia Post Office, Amcal Chemist, Guest's Café, Paul the Shoeman, Angelos, Fantasy Donuts, Snowy River Meats. Dapto Mall is on Moombara Street in Dapto CBD.

Redevelopment

The Mall went through huge redevelopments in early 2007. Alongside the Mall, a large car park was been built and was opened to the public on the 12th of February 2007. Stage one of the redevelopments opened on the 6th September 2007; the opening was planned for the 30th of October 2007, but then got pushed to the 31st. The Mall now boasts a food-court, Big W, 80 new specialty shops such as EB Games, Dick Smith Electronics, Beach Street Surf Scene, and many more. The redevelopments upgraded the Mall's total retail floor space from 6,000m²[5] to over 23,000m²[6] , making it the fourth-largest shopping centre in Wollongong, after competitors Westfield Warrawong, Stockland Shellharbour, and Wollongong Central.

Inside Dapto Mall, going up to level 1 via Travelator.

Dapto Ribbonwood Centre

The centre was Dapto's new community centre after relocating in 2001, it is named so after the species of tree that is planted in front. The local library is located on topmost floor of Level 1.

New, alternate entrance through Dapto Square. Note, this new section occupies the former western part of Byamee street.

Alternate view of the new alternate entrance through a Lane way

The northern facade

The view from the interior of the library on Level 1

Dapto Anglican Church

The **Dapto Anglican Church** centre is situated adjacent to the Dapto Mall and the Dapto Grayhound Racing Club; it also has services running from three other sites including the *St Aidens* building, the hall of *Dapto Primary School* on a Sunday morning, and the *St Lukes Brownsville* building (which is historically listed and over 100 years old). There are six services held weekly, as well as children and youth activities on Friday nights, children's programs on a Sunday morning and during the weeks of the school holidays, 'Cafe Church' during the week, and many one-off activities.

Dapto Uniting Church

Dapto Uniting Church is located in the heart of Dapto, next to the Dapto Medical Centre on the Princes Highway. Services are held at **9:30am** every Sunday morning and incorporate both contemporary and tradition styles of worship. **Children's programs** are held during the Sunday service, **Sunday's Cool**, and mid-week on Wednesdays, **Tiddlywinks playgroup**. The Dapto Uniting congregation is a healthy community composed of people of all generations, cultures and backgrounds.

Dapto Uniting Church.

The Dapto Uniting Church traces its roots right back to 1841 when Sunday services were first held in a Dapto schoolroom under the leadership of Presbyterian minister Rev. John Tait. With the establishment of the railway network in 1887, the Dapto area experienced a period of continued urban expansion over the ensuing decades. The current red brick building, St Andrews, was officially opened and dedicated on February 7, 1959. Construction work commenced in 1981 to attach new hall facilities to the existing church building. As they have now outgrown their current facilities, the congregation is hoping to relocate their church community to a brand new, more accessible site in the coming years.

For many years, the Dapto Uniting family has faithfully sought to serve the local community, through ministries such as the**Op Shop**, regular **Market Days, Showground Market Stall, Family Fun Days** and **Carols on the Lawn.**

St. John's Catholic Church

St. John's Church is the public worship centre of Dapto's Catholic population, and can be found in Jeramatta Street. Mass times are: Saturday Vigil Mass - 6:00pm, and Sundays at 07:30, 09:30 and 6:00pm. On the last Sunday of the month, cups of tea may be had after the 9:30 mass. It also has a new Youth Group.

See also

- Dapto High School
- Dapto Railway Station

References

[1] Australian Bureau of Statistics (25 October 2007). "Dapto (State Suburb)" (http://www.censusdata.abs.gov.au/ABSNavigation/prenav/
     LocationSearch?collection=Census&period=2006&areacode=SSC15086&producttype=QuickStats&breadcrumb=PL&action=401). *2006
     Census QuickStats.* . Retrieved 2008-11-20.
[2] "PLACE NAMES." (http://nla.gov.au/nla.news-article55185386). *The Australian Women's Weekly (1932-1982)* (1932-1982: National
     Library of Australia): p. 61. 13 May 1964. . Retrieved 22 February 2011.
[3] *New South Wales Private Railways - The Dapto District* Singleton, C.C. Australian Railway Historical Society Bulletin September, 1941
     pp37-39
[4] http://www.daptodogs.org.au/
[5] (http://www.wollongong.nsw.gov.au/development/planningforthefuture/Documents/Wollongong Retail Centres Study.pdf)
     Wollongong Retail Centre Study
[6] Dapto Mall Redevelopment (http://www.laingorourke.com.au/projects/building/commercial/project:dapto-mall-redevelopment)

External links

- Wollongong City Library - Dapto Local Statistics (http://www.wollongong.nsw.gov.au/library/localinfo/stats.html)
- Dapto Mall Website (http://www.daptomall.com.au/)
- Wollongong City Library - Dapto History (http://www.wollongong.nsw.gov.au/library/localinfo/dapto/history.html)
- Dapto Anglican Church (http://www.daptoanglican.org.au/)
- Dapto Uniting Church (http://www.dapto.unitingchurch.org.au/)
- Satellite photos from Google Maps (http://maps.google.com/maps?ll=-34.498705,150.794649&spn=0.033331,0.054957&t=k&hl=en)
- Dapto History (http://www.brucelloyd.net/history.htm)

Port Kembla, New South Wales

<table>
<tr><td colspan="2" align="center">Port Kembla
Wollongong, New South Wales</td></tr>
<tr><td colspan="2" align="center">View of Port Kembla from Hill 60 Park looking North West at Sunset</td></tr>
<tr><td>Population:</td><td>4,369[1]</td></tr>
<tr><td>Postcode:</td><td>2505</td></tr>
<tr><td>Coordinates:</td><td>34°28′S 150°54′E</td></tr>
<tr><td>Time zone: • Summer (DST)</td><td>AEST (UTC+10) AEDT (UTC+11)</td></tr>
<tr><td>Location:</td><td>112 km (70 mi) S of Sydney</td></tr>
<tr><td>LGA:</td><td>City of Wollongong</td></tr>
<tr><td>State District:</td><td>Wollongong</td></tr>
<tr><td>Federal Division:</td><td>Throsby</td></tr>
</table>

Suburbs around Port Kembla:		
Cringila	Coniston	Pacific Ocean
Warrawong	**Port Kembla**	Five Islands Nature Reserve
Kemblawarra	Primbee	Pacific Ocean

Port Kembla is a suburb of Wollongong 8 km south of the CBD and part of the Illawarra region of New South Wales. The suburb comprises a seaport, industrial complex (one of the largest in Australia), a small harbour foreshore nature reserve, and a small commercial sector. It is situated on the tip of Red Point, first sighted by Captain James Cook in 1770. The name "Kembla" is Aboriginal word meaning "plenty wildfowl".[2]

History

Before Port Kembla was an industrial suburb of Wollongong, it was a town with a remarkably self-sufficient society, a growing commercial centre, and a vibrant civic life. Town subdivision began in 1908, and by 1921 there were 1622 residents.[3] Economic expansion propelled further population growth.

Industrial Change

A new copper smelter and refinery, the Electrolytic Refinery and Smelting Company of Australia, began production in 1908, followed by the opening of Metal Manufactures in 1917 and finally the arrival of the Hoskins Iron and Steel Works in 1927, which became Australian Iron and Steel in 1928. By 1947 the town's population has increased to 4,960 with smaller satellite suburbs such as Cringila and Lake Heights, mushrooming on its fringes.[4] That year, 1947, marked the climax of a local campaign for municipal autonomy which was ultimately thwarted by the creation of a Greater City of Wollongong. In the post second world war period there was an inexorable decline of a 'Port

Kembla' society as local town boundaries were slowly but surely absorbed into a more Wollongong-focused or regional identity.[5]

Multiculturalism

Despite the decline from the heyday of the 1920s, the town experienced major social and demographic change in the 1950s and 1960s. Waves of migrants, mostly from the United Kingdom, Italy, Macedonia and Germany, moved to the town. During this period, Port Kembla was on the cusp on changes affecting Australian society generally as new ethic and cultural influences found a place in local society.[6] With its long migration history accommodating waves of migrant workers and their families, Port Kembla is still one of the most culturally diverse suburbs in New South Wales.

Hill 60

Port Kembla's highest point, Hill 60, overlooks the Five Islands and Red Point. Hill 60, originally the site of an Aboriginal settlement, was used by the army during World War II to make a coastal gun emplacement known as Illowra Battery. In September 1942, Aboriginal inhabitants were forcibly evicted from the area.[7] It has remained in the army's ownership and is now a public lookout reserve, despite a vigorous campaign to return some of the land to its Aboriginal owners.[8]

Dated: 1944 No. 2 gun Illowra Battery, Showing it's original BL 6 inch Mk XI gun

Gun position No. 2 at Illowra Battery, which formed part of the Kembla Fortress defences in World War II

Industry and Infrastructure

Port Kembla is known for the BlueScope Steel steelworks operations on Springhill Road and throughout North Port Kembla. Other notable industrial operations in the suburb are: Port Kembla Coal Terminal, Port Kembla Copper, Port Kembla Port Corporation, Incitec, Adstream Services, Port Kembla Gateway and Graincorp.

Rail

Port Kembla has a CityRail railway station on the Port Kembla branch of the South Coast railway line. It is the terminus of the branch line, and serves the residential area of the suburb of the same name. The station has one side platform, used for terminating trains. It is served by approximately one train per hour, usually a local service to Thirroul, but extra direct trains to and from Sydney are provided in the peak hours.

Pacific National operates daily coal trains to the Inner Harbour section of the port, and into the blast furnace section of the steelworks. Downer Rail has a workshop opposite the CityRail passenger terminus that services diesel powered locomotives for Pacific National.

Port

Port Kembla Harbour is a major export location for coal mined in the southern and western regions of New South Wales. As part of the state governments plan to divert ships containing auto mobiles, the port has received significant upgrades and infrastructure including a new Maritime Office and many jobs have been created as the need for port logistics grows. Patrick Corporation holds a contract for integrated port services in the harbour and transports goods by road or rail through its parent company Pacific National.

Port Kembla Harbour, taken from Breakwater Battery

Sports and Leisure

Sporting Teams

Port Kembla has both junior and senior teams in local popular sporting leagues such as:

- Port Kembla Rugby League in the Illawarra Division Rugby League

Home grounds are Noel Mulligan Oval[9]

- Port Kembla Cricket Club in the Cricket Illawarra Competition

Home grounds are King George V Park[10]

- Port Kembla Soccer Club in the Illawarra Football Association

Home grounds are Darcy Wentworth Park[11]

- Port Kembla AFL Club [12] plays in South Coast AFL and AFL Illawarra

Home grounds are Kully Bay Park[13]

Port Kembla Rugby League, Port Kembla Soccer Club and Port Kembla AFL home grounds are not located in Port Kembla, they all play in parks across Warrawong.

Parks and Beaches

Port Kembla has a number of parks, nature reserves, beaches and a Saltwater Olympic pool:

- King George V Park[14]

A foreshore park located in walking distance from Port Kembla Beach. Used in summer for Port Kembla Cricket Club home games.

- Hill 60 Park[15]

A popular take off area for hang gliders and para gliders, Hill 60 Park has BBQ facilities as well as picnic shelters, seats and tables positioned to enjoy the scenic views.

- Port Kembla Beach[16]

An award winning beach[17] , seasonally patrolled from September to April[18] and home to the Port Kembla Surf Life Saving Club.

- Fishermans Beach[19]

A small sheltered beach at the bottom of Hill 60's eastern side facing the Five Islands Nature Reserve.

- North Port Kembla Beach[20]

Also known by locals as MM Beach for is close proximity to the Metal Manufacturers site on Gloucester Boulevard. Remenants of a tidal rock pool are still standing near the southern end of the beach below the headland.

Public transport

Train

Further information: Port Kembla railway station, New South Wales, Port Kembla North railway station, New South Wales, South Coast railway line.

Port Kembla has two railway stations. Port Kembla railway station is a CityRail railway station on the Port Kembla branch of the South Coast railway line. It is the terminus of the branch line, and serves the residential area of Port Kembla.

Port Kembla North is a CityRail railway station on the Port Kembla branch of the South Coast railway line. It serves the industrial area of the suburb of Port Kembla to the south-east of Wollongong. The station is the nearest to the BHP Billiton site in the area.

Both stations have a one sided platform, with the platform at Port Kembla used for terminating trains. The stations are served by approximately one train per hour, usually a local service to Thirroul, but extra direct trains to and from Sydney are provided in the peak hours.

Port Kembla Station

Port Kembla North Station

Bus

Premier Illawarra runs three routes to and from Port Kembla railway station[21] :

- 34 - to Wollongong via Berkeley
- 43 - to Dapto
- 65 - to North Beach

Health and Environmental Issues

Port Kembla Chimney Stack

Port Kembla is home to one of Australia's tallest industrial chimneys, a 198 metre tall chimney built in 1965. Port Kembla Primary School was once located adjacent to it but was closed down due to pollution problems from the chimney including lead contaminated soil, acid rain and soot. A warning alarm was fitted to warn of high toxin levels. In November 2008, the Port Kembla stack was inspected and confirmed to have concrete cancer.[22] The stack was planned to be demolished in early 2010 at a cost of A$10 million.

[23] As of 6th September 2010 the plans to knock down the stack have been revised by the NSW Department of Planning. These plans include demolition of the existing Port Kembla Copper structures surrounding the chimney, excluding the locally heritage listed Precious Metals Mill Chimney and the Assay Offices. The work is now due to start in the middle of 2011 with a team of 30 workers, under supervision by NSW Police, NSW WorkCover and relevant emergency services at a cost of A$8 million with an expected time frame of 16 months.

Industrial pollution

Port Kembla's industrial heart has caused significant environmental problems. The heavy industry of the area makes significant emissions of nitrogen oxides and other dangerous gases. Gas emissions also result in the formation of acid rain which has caused spotting of metal surfaces and the rapid corrosion of many structures to be reported by locals and other residents in neighbouring suburbs.[24]

Aerial photo from north west

Health problems associated with noxious gases are more common. One 1998 study of the industrial areas of Newcastle and Port Kembla found 'an important association between relatively low levels of particulate air pollution and respiratory symptoms' among primary school children.[25] Fallout has also introduced elevated levels of lead and other heavy metals to the soil around Port Kembla and has formed thick deposits in many buildings and industrial structures.[26]

See also

- BlueScope Steel
- Port Kembla Port Corporation

References

[1] Australian Bureau of Statistics (25 October 2007). "Port Kembla (State Suburb)" (http://www.censusdata.abs.gov.au/ABSNavigation/
prenav/LocationSearch?collection=Census&period=2006&areacode=SSC15211&producttype=QuickStats&breadcrumb=PL&
action=401). *2006 Census QuickStats*. . Retrieved 2009-01-14.

[2] "PLACE NAMES." (http://nla.gov.au/nla.news-article55185386). *The Australian Women's Weekly (1932-1982)* (1932-1982: National
Library of Australia): p. 61. 13 May 1964. . Retrieved 22 February 2011.

[3] Commonwealth Census of Australia, 1921

[4] Commonwealth Census of Australia, 1947

[5] See Erik Eklund, Steel Town: the making and breaking of Port Kembla, pp.131-136

[6] Erik Eklund, Steel Town: the making and breaking of Port Kembla, pp. 158-171

[7] See Erik Eklund, 'Steel Town: the making and breaking of Port Kembla', MUP, Melbourne, 2002, pp. 114-130

[8] http://www.michaelorgan.org.au/hill60.htm

[9] http://maps.google.com.au/maps/ms?ie=UTF8&source=embed&oe=UTF8&msa=0&msid=106987883233634634262.
0004856280c83d740b1d5

[10] http://maps.google.com/maps/place?num=20&hl=en&safe=off&um=1&ie=UTF-8&q=king+george+v+park+port+kembla&
fb=1&hq=king+george+v+park+port+kembla&hnear=king+george+v+park+port+kembla&cid=1081283275369116352

[11] http://maps.google.com.au/maps/ms?ie=UTF8&source=embed&oe=UTF8&msa=0&msid=106987883233634634262.
0004856280c83d740b1d5

[12] http://www.pkafc.asn.au/

[13] http://maps.google.com/maps/place?cid=1081283275369116384&q=Kully+Bay+Park,+Wollongong,+New+South+Wales,+
Australia&hl=en&sll=-34.489383,150.886379&sspn=0.022072,0.038418&ie=UTF8&ll=-34.468849,150.840998&spn=0,0&z=14

[14] http://maps.google.com/maps/place?cid=1081283275369116352&q=King+George+V+Park+port&hl=en&sll=-34.489087,150.
909646&sspn=0.022072,0.038418&ie=UTF8&ll=-34.467151,150.864258&spn=0,0&z=14

[15] http://maps.google.com/maps/place?cid=1081283275369116304&q=hill+60+Park+port&hl=en&sll=-34.489107,150.915826&
sspn=0.022072,0.038418&ie=UTF8&ll=-34.467151,150.870438&spn=0,0&z=14

[16] http://maps.google.com/maps/place?ftid=0x6b131706f8ce8a4d:0x4701e95e76169ac0&q=port+kembla+beach&hl=en&sll=-34.
495866,150.903848&sspn=0.018393,0.032015&ie=UTF8&ll=-34.484911,150.881166&spn=0,0&z=15

[17] http://www.illawarramercury.com.au/news/local/news/general/its-official-port-kemblas-the-best-beach-in-nsw/1696569.aspx

[18] http://www.wollongong.nsw.gov.au/facilities/beachespools/Pages/beaches.aspx

[19] http://maps.google.com/maps/place?ftid=0x6b1317a33cef0099:0xb415a77febf9d76b&q=fishermans+beach+port+kembla&hl=en&
sll=-34.489977,150.917029&sspn=0.018394,0.032015&ie=UTF8&ll=-34.479003,150.894341&spn=0,0&z=15

[20] http://maps.google.com/maps/place?ftid=0x6b13179bfb517b33:0x75fa716bbf19a176&q=north+beach+port+kembla&hl=en&
sll=-34.480407,150.913457&sspn=0.018397,0.032015&ie=UTF8&ll=-34.469451,150.890737&spn=0,0&z=15

[21] http://premierillawarra.com.au/timetables.html#berkeley

[22] Illawarra Mercury, 26th November 2008

[23] http://www.planning.nsw.gov.au/DesktopModules/MediaCentre/getdocument.aspx?mid=449

[24] http://ro.uow.edu.au/cgi/viewcontent.cgi?article=1014&context=wollgeo

[25] Peter R Lewis et al, (1998) 'Outdoor air pollution and children's respiratory symptoms in the steel cities of New South Wales', *Medical
Journal of Australia*, 169: pp. 459-463 See (http://www.mja.com.au/public/issues/nov2/lewis/lewis.html)

[26] http://www.lead.org.au/lanv5n1/lanv5n1-5.html

External links

- Australia's Tallest Industrial Chimney - Sightseeing with Google Satellite Maps (http://www.satellite-sightseer.
com/id/10301/Australia/New_South_Wales/Wollongong/Australias_Tallest_Industrial_Chimney)

Article Sources and Contributors

Albion Park railway station *Source*: http://en.wikipedia.org/w/index.php?title=Albion_Park_railway_station *Contributors*: Alex.Mcgregor, Crusoe8181, Da monster under your bed, GrahamHardy, Grahamec, Harryboyles, Hendikins, Jpp42, MrHarper, Quaidy, Sb617, SimonP, Slambo, Stepheng3, The Fulch, Waacstats, Wrelwser43, 10 anonymous edits

CityRail *Source*: http://en.wikipedia.org/w/index.php?title=CityRail *Contributors*: 1717, 99of9, Aaronsclee, Aldy, Alex.Mcgregor, Alex.tan, Aloha x, Andrew.bosa, André Devecserii, Angr, Ansend, Ansett, Ashill, AtholM, Aucitypops, Backslash Forwardslash, Ben Ben, Bendiviolet, Beneaththelandslide, Bhludzin, Bidgee, Bja1608, Bobrayner, Burntsauce, Cahill1, Camembert, Cdlw93, ChampagneComedy, Chanlord, Chris the speller, Clarkk, Coelacan, Coolkids000, Crusoe8181, DabMachine, Damaster98, Darkcore, David Edgar, Dbfirs, DdInfiltrator, DirectEdge, Djk3, Djvera, DrFrench, Dysprosia, Earthahead, Eliz2009, Endarrt, Enochlau, Eraserhead1, Essolo, Eternal dragon, Euryalus, F Notebook, Factrules, Farosdaughter, Farras Octara, Fastily, Felix Portier, Fetchcomms, FetchcommsAWB, Finlay McWalter, Firehose, Fryed-peach, Gareth Aus, Goninan bl00d, Gonzonoir, Grahamec, HarlandQPitt, Harriseldon, Harryboyles, Hirohisat, Hornetfig, Humehwy, Hurstville1, Ianblair23, Into The Fray, Invenio, Ioexplore, Iten, Iwishihadalife, JPD, JRG, Jmw2508, Johnbevan, Joy, Just Another Dan, Kewpid, Kransky, Lacrimosus, LedgendGamer, Lightmouse, Llamad123, MER-C, Markaci, MasterMind5991, Matt037291, Maxim75, Michaeldragon800, Mikecraig, Miq, MoondyneAWB, Mw12310, Mysid, Nalini sharma1984, Nomadtales, Ojw, Oknazevad, One Salient Oversight, OpenToppedBus, Orderinchaos 2, Osarius, Oxymoron83, PDH, Patrick, Pgan002, Phlyght, Picapica, Pickles8, Pitaman, Plastikspork, Playclever, Poplar22, Purkaeus, Quaidy, Qwerty Binary, Qxz, ROBSKE, Randwicked, Rebecca, Recurring dreams, Rich Farmbrough, Richyhan, Ricky81682, Rjwilmsi, Rkurzawa, Robert Brockway, Roke, SDC, SM247, SPUI, SSR600, Sam67fr, Sebsf, Shadowjams, Shy Guy Gunzel, Single16+Sections, Slambo, Sliggy, Snowfreq, Somebody in the WWW, Spiky Sharkie, StaticSan, Stephen Bain, Steven Fitter, StraussianNeocon, Subsolar, Sumple, SuperMagicRabbit, Susurrus, Syd1435, Sydney.city.easts, T4717, TPK, Tabletop, Tangara21, TbohlsenNSWSSMRC, Tbsoftware, Techelf, Thadocta, The JPS, Thingg, Thortful, Todd661, Traintro, Trjumpet, Troyon, Voyager, WMXX, Wcp07, Werdna, Wittylama, Wongm, Woohookitty, Wprlh, Wykymania, YuMaNuMa, Zerolimits, Zigger, Zoicon5, Zro, , 373 anonymous edits

South Coast railway line, New South Wales *Source*: http://en.wikipedia.org/w/index.php?title=South_Coast_railway_line%2C_New_South_Wales *Contributors*: Axpde, Clarkk, Commissioner Geoff, Crusoe8181, Djsasso, Drpurpleturtle, Dysprosia, E=MC^2, Gareth Aus, Gonzerelli, Gorillazfeelgoodinc, Grahamec, Grogan deYobbo, Harryboyles, Ianblair23, JRG, Joseph Solis in Australia, Ka-ru, Karl Dickman, Maebmij, Maias, Mhiji, Michael Glass, MikeyAus069, MisfitToys, Nomadtales, Powerfuel, Quaidy, Qualoga, Rich Farmbrough, Richyhan, Rjwilmsi, Single16+Sections, Somebody in the WWW, Spiky Sharkie, Tabletop, The Fulch, Thryduulf, Traintro, Wprlh, 13 anonymous edits

Albion Park Rail, New South Wales *Source*: http://en.wikipedia.org/w/index.php?title=Albion_Park_Rail%2C_New_South_Wales *Contributors*: AussieLegend, Bidgee, Borofkin, CarolGray, Crusoe8181, Emersoni, Frankie816, Graeme Bartlett, Grahamec, Grogan deYobbo, Jayden Powell, Jpp42, Kekman64, LeoNomis, Longhair, LordVetinari, Nath1991, Orderinchaos, ScottDavis, SimonP, Xposya, 35 anonymous edits

Albion Park, New South Wales *Source*: http://en.wikipedia.org/w/index.php?title=Albion_Park%2C_New_South_Wales *Contributors*: Batobatobato, Bidgee, Bocden, Borofkin, Boydy83, CarolGray, Chris the speller, Crusoe8181, Eccy89, Emersoni, Emily Kate Cullen Jones, Fawcett5, Frankie816, Gaius Cornelius, Graeme Bartlett, Grahamec, Grogan deYobbo, Ianblair23, Insanity Incarnate, J.delanoy, Jayden Powell, Kappa, Kekman64, LilHelpa, Lilyth, Longhair, LordVetinari, Mattparko, Merbabu, Mindmatrix, Mmxx, Nationalfooty, Noreset, Orderinchaos, Pax85, Quaidy, Reader781, Rileyb, Sasquatch, ScottDavis, Sejtraav, SimonP, Sjakkalle, Thatguythere570, Tijuana Brass, Tony1, VirtualSteve, 66 anonymous edits

Oak Flats railway station *Source*: http://en.wikipedia.org/w/index.php?title=Oak_Flats_railway_station *Contributors*: Alex.Mcgregor, Badagnani, Crusoe8181, Da monster under your bed, Ezzaryn, Gareth Aus, GrahamHardy, Grahamec, Harryboyles, Hendikins, MrHarper, Sb617, Spiky Sharkie, The Fulch, 5 anonymous edits

Dapto railway station *Source*: http://en.wikipedia.org/w/index.php?title=Dapto_railway_station *Contributors*: Abesty, Aeonx, Alex.Mcgregor, Badagnani, Crusoe8181, Da monster under your bed, Endarrt, Gonzerelli, GrahamHardy, Grahamec, Harryboardman, Harryboyles, Hendikins, JRG, MrHarper, Naylor101, Sb617, SimonP, Somebody in the WWW, The Fulch, Waacstats, Wykymania, 20 anonymous edits

Bomaderry railway station *Source*: http://en.wikipedia.org/w/index.php?title=Bomaderry_railway_station *Contributors*: Alex.Mcgregor, Bleakcomb, Crusoe8181, DJGB, Da monster under your bed, GrahamHardy, Grahamec, Harryboyles, Hendikins, JRG, MrHarper, Norry888, PoccilScript, Quaidy, Sb617, The Fulch, Utcursch, 6 anonymous edits

Shellharbour, New South Wales *Source*: http://en.wikipedia.org/w/index.php?title=Shellharbour%2C_New_South_Wales *Contributors*: Academic Challenger, Andrewag, Ataafe, Ausref, Bidgee, Blarneytherinosaur, Borofkin, Breno, Chris the speller, Clarkk, Ctjf83, Donarreiskoffer, Gene Nygaard, Grahamec, Grogan deYobbo, Harryboyles, Ianblair23, JRG, Jayden Powell, Longhair, LordVetinari, Night of the Big Wind Turbo, Rebecca, Scharks, Seb26, ShellyWorkers07, Shellyboy, Sumple, Surferpink, The celt, WheelieBinner, William Graham, Xposya, 48 anonymous edits

Wollongong railway station *Source*: http://en.wikipedia.org/w/index.php?title=Wollongong_railway_station *Contributors*: Aeonx, Alex.Mcgregor, Cdlw93, Crusoe8181, Da monster under your bed, Endarrt, GrahamHardy, Grahamec, Harryboyles, Hendikins, JRG, MrHarper, Sb617, Spiky Sharkie, The Fulch, 7 anonymous edits

Dapto, New South Wales *Source*: http://en.wikipedia.org/w/index.php?title=Dapto%2C_New_South_Wales *Contributors*: 99of9, Adam.J.W.C., Amren, Andrewtss, Atomius, Auhsor, Barticus88, Bidgee, Bocden, Brenont, Camerong, Ckmeow, Clarkk, Commissioner Geoff, Dapto@mailinator.com, Dawnseeker2000, Dsp13, Eliz81, Eregli bob, EurekaLott, Everyking, Evil saltine, Fatkid21, Firsfron, Gonzo fan2007, Graeme Bartlett, Graham87, Grahamec, Grogan deYobbo, Guard952, Hadal, Insanity Incarnate, Iridescent, Jm1234567890, Joelster, John Vandenberg, Jsann, Kerry Raymond, LAX, Lollipiop, Longhair, LordVetinari, M youngy, Madchester, Martarius, Matthewrbowker, Mattinbgn, Mattv1.0, Meluvseveryone, Michael Devore, Mindmatrix, Moolahman, Nezzadar, Nickquinnmadlad, No doid, Olnalra, Orderinchaos, Ottawa4ever, PeaceNT, Pi, Psmith, Quarl, Recurring dreams, Rjwilmsi, Rmosler2100, Rom rulz424, Saga City, SaveThePoint, ScottDavis, Somebody in the WWW, Steven Fitter, T.J.V., Thatguythere570, Thortful, Tijuana Brass, Torchwoodwho, Triwbe, VegitaU, Viriditas, Westogent, Wykymania, 152 anonymous edits

Port Kembla, New South Wales *Source*: http://en.wikipedia.org/w/index.php?title=Port_Kembla%2C_New_South_Wales *Contributors*: Adam.J.W.C., Angusmclellan, Atomius, BenAveling, Biasia, Bidgee, Borofkin, CRKingston, Clarkk, Drpickem, Eekconsummer, Euryalus, Fawcett5, Firsfron, Francisco Del Piero, Fusionfusion, Gillyweed, Graeme Bartlett, Grahamec, Grant65, Grogan deYobbo, Harryboyles, Hmains, JRG, Jayden Powell, Jengod, Kerry Raymond, Kocovski, Laager, Lancsalot, Lissajous, Longhair, LordVetinari, Martarius, Mattinbgn, Mindmatrix, Miracle Pen, MoondyneAWB, Nationalfooty, Neschek, Newm30, Orestes654, Pk2505, PoorPhotoremovalist, Portk, Qwerty Binary, RJFJR, Rcbutcher, Recurring dreams, Saberwyn, Scorpiamdj, ScottDavis, Stepheng3, The Thing That Should Not Be, Wyp, 51 anonymous edits

Image Sources, Licenses and Contributors

Image:Cityrailsign.svg *Source*: http://en.wikipedia.org/w/index.php?title=File:Cityrailsign.svg *License*: Public Domain *Contributors*: Original uploader was Bilious at en.wikipedia

Image:Albion Park.jpg *Source*: http://en.wikipedia.org/w/index.php?title=File:Albion_Park.jpg *License*: Creative Commons Attribution-Sharealike 2.5 *Contributors*: User:Grahamec

Image:Wheelchair symbol.svg *Source*: http://en.wikipedia.org/w/index.php?title=File:Wheelchair_symbol.svg *License*: Public Domain *Contributors*: ALE!, David Levy, Dream out loud, Ktims, Lensovet, NE2, Sarang, Stifle, Wst, 2 anonymous edits

File:CR Plat 1.png *Source*: http://en.wikipedia.org/w/index.php?title=File:CR_Plat_1.png *License*: GNU Free Documentation License *Contributors*: User:Endarrt

File:CR Plat 2.png *Source*: http://en.wikipedia.org/w/index.php?title=File:CR_Plat_2.png *License*: GNU Free Documentation License *Contributors*: User:Endarrt

File:CityRail new logo.png *Source*: http://en.wikipedia.org/w/index.php?title=File:CityRail_new_logo.png *License*: unknown *Contributors*: Troyon

File:SydneyMuseumStation2crop gobeirne.jpg *Source*: http://en.wikipedia.org/w/index.php?title=File:SydneyMuseumStation2crop_gobeirne.jpg *License*: Creative Commons Attribution 2.5 *Contributors*: User:gobeirne

Image:Inside central railway station, sydney.jpg *Source*: http://en.wikipedia.org/w/index.php?title=File:Inside_central_railway_station,_sydney.jpg *License*: Creative Commons Attribution-Sharealike 2.5 *Contributors*: 17177, Gareth, Voyager

Image:A set end carriage cityrail.jpg *Source*: http://en.wikipedia.org/w/index.php?title=File:A_set_end_carriage_cityrail.jpg *License*: Creative Commons Attribution-Sharealike 3.0 *Contributors*: User:Wykymania

Image:Cityrail-millennium-M32-ext.jpg *Source*: http://en.wikipedia.org/w/index.php?title=File:Cityrail-millennium-M32-ext.jpg *License*: Creative Commons Attribution 2.5 *Contributors*: User:Alexanderino

File:Cityrail ticket.jpg *Source*: http://en.wikipedia.org/w/index.php?title=File:Cityrail_ticket.jpg *License*: Public Domain *Contributors*: User:Maxim75

Image:Sydney railway map.gif *Source*: http://en.wikipedia.org/w/index.php?title=File:Sydney_railway_map.gif *License*: Creative Commons Attribution 3.0 *Contributors*: John Shadbolt Original uploader was JohnnoShadbolt at en.wikipedia

Image:CityRail InterCity map.png *Source*: http://en.wikipedia.org/w/index.php?title=File:CityRail_InterCity_map.png *License*: Public Domain *Contributors*: Essolo

Image:Artarmon railway station Sydney platform.jpg *Source*: http://en.wikipedia.org/w/index.php?title=File:Artarmon_railway_station_Sydney_platform.jpg *License*: GNU Free Documentation License *Contributors*: Gareth, Grahamec, JROBBO, Quickstep, Voyager

File:Epping station underground platform.JPG *Source*: http://en.wikipedia.org/w/index.php?title=File:Epping_station_underground_platform.JPG *License*: Creative Commons Attribution-Sharealike 3.0 *Contributors*: User:Gareth

Image:Dual_screen_cityrail_board.jpg *Source*: http://en.wikipedia.org/w/index.php?title=File:Dual_screen_cityrail_board.jpg *License*: Creative Commons Attribution-Sharealike 3.0 *Contributors*: User:Wykymania

Image:Cityrail concourse outer suburban indicator.jpg *Source*: http://en.wikipedia.org/w/index.php?title=File:Cityrail_concourse_outer_suburban_indicator.jpg *License*: Creative Commons Attribution-Sharealike 3.0 *Contributors*: User:Wykymania

Image:CityRail Indicator Boards.JPG *Source*: http://en.wikipedia.org/w/index.php?title=File:CityRail_Indicator_Boards.JPG *License*: Creative Commons Attribution-Sharealike 3.0 *Contributors*: User:Wykymania

Image:Inner city cityrail board.jpg *Source*: http://en.wikipedia.org/w/index.php?title=File:Inner_city_cityrail_board.jpg *License*: Creative Commons Attribution-Sharealike 3.0 *Contributors*: User:Wykymania

Image:Cityrail next train to city indicator .jpg *Source*: http://en.wikipedia.org/w/index.php?title=File:Cityrail_next_train_to_city_indicator_.jpg *License*: Creative Commons Attribution-Sharealike 3.0 *Contributors*: User:Wykymania

Image:Cityrail next trains to city.jpg *Source*: http://en.wikipedia.org/w/index.php?title=File:Cityrail_next_trains_to_city.jpg *License*: Creative Commons Attribution-Sharealike 3.0 *Contributors*: User:Wykymania

Image:Cityrail intercity interstate indicator.jpg *Source*: http://en.wikipedia.org/w/index.php?title=File:Cityrail_intercity_interstate_indicator.jpg *License*: Creative Commons Attribution-Sharealike 3.0 *Contributors*: User:Wykymania

Image:Bradfield Scheme Sydney CBD Railways alt.png *Source*: http://en.wikipedia.org/w/index.php?title=File:Bradfield_Scheme_Sydney_CBD_Railways_alt.png *License*: unknown *Contributors*: Original uploader was Hornetfig at en.wikipedia

Image:CityRail yellow line.jpg *Source*: http://en.wikipedia.org/w/index.php?title=File:CityRail_yellow_line.jpg *License*: Creative Commons Attribution-Sharealike 2.5 *Contributors*: Enochlau, 1 anonymous edits

File:Help point cityrail.jpg *Source*: http://en.wikipedia.org/w/index.php?title=File:Help_point_cityrail.jpg *License*: Creative Commons Attribution-Sharealike 3.0 *Contributors*: User:Wykymania

Image:RonChristieplan.jpg *Source*: http://en.wikipedia.org/w/index.php?title=File:RonChristieplan.jpg *License*: Creative Commons Attribution-Sharealike 2.0 *Contributors*: JRG, Quadell

File:CityRailinterblue.png *Source*: http://en.wikipedia.org/w/index.php?title=File:CityRailinterblue.png *License*: Public Domain *Contributors*: Original image released to public domain by Single16+Sections. Converted to PNG by Gareth.

Image:Como Bridge.JPG *Source*: http://en.wikipedia.org/w/index.php?title=File:Como_Bridge.JPG *License*: GNU Free Documentation License *Contributors*: J Bar

Image:Coledale railway station.2005-07-06.jpg *Source*: http://en.wikipedia.org/w/index.php?title=File:Coledale_railway_station.2005-07-06.jpg *License*: Creative Commons Attribution-Sharealike 2.1 *Contributors*: samwilson

Image:H11-scarborough.jpg *Source*: http://en.wikipedia.org/w/index.php?title=File:H11-scarborough.jpg *License*: Creative Commons Attribution-Sharealike 3.0 *Contributors*: Somebody in the WWW

Image:L2-cringila.jpg *Source*: http://en.wikipedia.org/w/index.php?title=File:L2-cringila.jpg *License*: Creative Commons Attribution-Sharealike 3.0 *Contributors*: Somebody in the WWW

Image:Wollongong station.jpg *Source*: http://en.wikipedia.org/w/index.php?title=File:Wollongong_station.jpg *License*: Creative Commons Attribution-Sharealike 3.0 *Contributors*: User:Grahamec

File:Albion NSW Aerial.JPG *Source*: http://en.wikipedia.org/w/index.php?title=File:Albion_NSW_Aerial.JPG *License*: Creative Commons Attribution 3.0 *Contributors*: User:Graeme Bartlett

File:Albion_Park_aerial.jpg *Source*: http://en.wikipedia.org/w/index.php?title=File:Albion_Park_aerial.jpg *License*: Creative Commons Attribution-Sharealike 3.0 *Contributors*: User:Graeme Bartlett

Image:Oak Flats station 2.jpg *Source*: http://en.wikipedia.org/w/index.php?title=File:Oak_Flats_station_2.jpg *License*: Creative Commons Attribution-Sharealike 3.0 *Contributors*: User:Grahamec

Image:Oak Flats station 1.jpg *Source*: http://en.wikipedia.org/w/index.php?title=File:Oak_Flats_station_1.jpg *License*: Creative Commons Attribution-Sharealike 3.0 *Contributors*: User:Grahamec

Image:Dapto station 1.jpg *Source*: http://en.wikipedia.org/w/index.php?title=File:Dapto_station_1.jpg *License*: Creative Commons Attribution-Sharealike 3.0 *Contributors*: User:Grahamec

File:CR Plat 3.png *Source*: http://en.wikipedia.org/w/index.php?title=File:CR_Plat_3.png *License*: GNU Free Documentation License *Contributors*: User:Endarrt

Image:Dapto station 2.jpg *Source*: http://en.wikipedia.org/w/index.php?title=File:Dapto_station_2.jpg *License*: Creative Commons Attribution-Sharealike 3.0 *Contributors*: User:Grahamec

Image:Dapto Station Plan.jpg *Source*: http://en.wikipedia.org/w/index.php?title=File:Dapto_Station_Plan.jpg *License*: Public Domain *Contributors*: Abesty, Wykimania

Image:Bombaderry Station.jpg *Source*: http://en.wikipedia.org/w/index.php?title=File:Bombaderry_Station.jpg *License*: Creative Commons Attribution-Sharealike 2.5 *Contributors*: User:Grahamec

File:Nowra Bridge.jpg *Source*: http://en.wikipedia.org/w/index.php?title=File:Nowra_Bridge.jpg *License*: Creative Commons Attribution-Sharealike 3.0 *Contributors*: User:Grahamec

File:Bombaderry railway bridge 1.jpg *Source*: http://en.wikipedia.org/w/index.php?title=File:Bombaderry_railway_bridge_1.jpg *License*: Creative Commons Attribution-Sharealike 2.5 *Contributors*: User:Grahamec

File:BomaderryNowraStation1.JPG *Source*: http://en.wikipedia.org/w/index.php?title=File:BomaderryNowraStation1.JPG *License*: unknown *Contributors*: DJGB

File:Sharbour1.JPG *Source*: http://en.wikipedia.org/w/index.php?title=File:Sharbour1.JPG *License*: Public Domain *Contributors*: Andrewag

Image:Wollongong station dock platform.jpg *Source*: http://en.wikipedia.org/w/index.php?title=File:Wollongong_station_dock_platform.jpg *License*: Creative Commons Attribution-Sharealike 3.0 *Contributors*: User:Grahamec

License

Creative Commons Attribution-ShareAlike 3.0 Unported - Deed

This is a human-readable summary of the Creative Commons Attribution ShareAlike 3.0 Unported License (http://en.wikipedia.org/wiki/Wikipedia:Text_of_Creative_Commons_Attribution-ShareAlike_3.0_Unported_License)
You are free:

- **to Share**—to copy, distribute and transmit the work, and
- **to Remix**—to adapt the work

Under the following conditions:

- **Attribution**—You must attribute the work in the manner specified by the author or licensor (but not in any way that suggests that they endorse you or your use of the work.)
- **Share Alike**—If you alter, transform, or build upon this work, you may distribute the resulting work only under the same, similar or a compatible license.

With the understanding that:

- **Waiver**—Any of the above conditions can be waived if you get permission from the copyright holder.
- **Other Rights**—In no way are any of the following rights affected by the license:
  - your fair dealing or fair use rights;
  - the author's moral rights; and
  - rights other persons may have either in the work itself or in how the work is used, such as publicity or privacy rights.
- **Notice**—For any reuse or distribution, you must make clear to others the license terms of this work. The best way to do that is with a link to http://creativecommons.org/licenses/by-sa/3.0/

GNU Free Documentation License

As of July 15, 2009 Wikipedia has moved to a dual-licensing system that supersedes the previous GFDL only licensing. In short, this means that text licensed under the GFDL can no longer be imported to Wikipedia. Additionally, text contributed after that date can not be exported under the GFDL license. See Wikipedia:Licensing update for further information.

Version 1.3, 3 November 2008 Copyright (C) 2000, 2001, 2002, 2007, 2008 Free Software Foundation, Inc. <http://fsf.org/>
Everyone is permitted to copy and distribute verbatim copies of this license document, but changing it is not allowed.

0. PREAMBLE

The purpose of this License is to make a manual, textbook, or other functional and useful document "free" in the sense of freedom: to assure everyone the effective freedom to copy and redistribute it, with or without modifying it, either commercially or noncommercially. Secondarily, this License preserves for the author and publisher a way to get credit for their work, while not being considered responsible for modifications made by others.
This License is a kind of "copyleft", which means that derivative works of the document must themselves be free in the same sense. It complements the GNU General Public License, which is a copyleft license designed for free software.
We have designed this License in order to use it for manuals for free software, because free software needs free documentation: a free program should come with manuals providing the same freedoms that the software does. But this License is not limited to software manuals; it can be used for any textual work, regardless of subject matter or whether it is published as a printed book. We recommend this License principally for works whose purpose is instruction or reference.

1. APPLICABILITY AND DEFINITIONS

This License applies to any manual or other work, in any medium, that contains a notice placed by the copyright holder saying it can be distributed under the terms of this License. Such a notice grants a world-wide, royalty-free license, unlimited in duration, to use that work under the conditions stated herein. The "Document", below, refers to any such manual or work. Any member of the public is a licensee, and is addressed as "you". You accept the license if you copy, modify or distribute the work in a way requiring permission under copyright law.
A "Modified Version" of the Document means any work containing the Document or a portion of it, either copied verbatim, or with modifications and/or translated into another language.
A "Secondary Section" is a named appendix or a front-matter section of the Document that deals exclusively with the relationship of the publishers or authors of the Document to the Document's overall subject (or to related matters) and contains nothing that could fall directly within that overall subject. (Thus, if the Document is in part a textbook of mathematics, a Secondary Section may not explain any mathematics.) The relationship could be a matter of historical connection with the subject or with related matters, or of legal, commercial, philosophical, ethical or political position regarding them.
The "Invariant Sections" are certain Secondary Sections whose titles are designated, as being those of Invariant Sections, in the notice that says that the Document is released under this License. If a section does not fit the above definition of Secondary then it is not allowed to be designated as Invariant. The Document may contain zero Invariant Sections. If the Document does not identify any Invariant Sections then there are none.
The "Cover Texts" are certain short passages of text that are listed, as Front-Cover Texts or Back-Cover Texts, in the notice that says that the Document is released under this License. A Front-Cover Text may be at most 5 words, and a Back-Cover Text may be at most 25 words.
A "Transparent" copy of the Document means a machine-readable copy, represented in a format whose specification is available to the general public, that is suitable for revising the document straightforwardly with generic text editors or (for images composed of pixels) generic paint programs or (for drawings) some widely available drawing editor, and that is suitable for input to text formatters or for automatic translation to a variety of formats suitable for input to text formatters. A copy made in an otherwise Transparent file format whose markup, or absence of markup, has been arranged to thwart or discourage subsequent modification by readers is not Transparent. An image format is not Transparent if used for any substantial amount of text. A copy that is not "Transparent" is called "Opaque".
Examples of suitable formats for Transparent copies include plain ASCII without markup, Texinfo input format, LaTeX input format, SGML or XML using a publicly available DTD, and standard-conforming simple HTML, PostScript or PDF designed for human modification. Examples of transparent image formats include PNG, XCF and JPG. Opaque formats include proprietary formats that can be read and edited only by proprietary word processors, SGML or XML for which the DTD and/or processing tools are not generally available, and the machine-generated HTML, PostScript or PDF produced by some word processors for output purposes only.
The "Title Page" means, for a printed book, the title page itself, plus such following pages as are needed to hold, legibly, the material this License requires to appear in the title page. For works in formats which do not have any title page as such, "Title Page" means the text near the most prominent appearance of the work's title, preceding the beginning of the body of the text.
The "publisher" means any person or entity that distributes copies of the Document to the public.
A section "Entitled XYZ" means a named subunit of the Document whose title either is precisely XYZ or contains XYZ in parentheses following text that translates XYZ in another language. (Here XYZ stands for a specific section name mentioned below, such as "Acknowledgements", "Dedications", "Endorsements", or "History".) To "Preserve the Title" of such a section when you modify the Document means that it remains a section "Entitled XYZ" according to this definition.
The Document may include Warranty Disclaimers next to the notice which states that this License applies to the Document. These Warranty Disclaimers are considered to be included by reference in this License, but only as regards disclaiming warranties: any other implication that these Warranty Disclaimers may have is void and has no effect on the meaning of this License.

2. VERBATIM COPYING

You may copy and distribute the Document in any medium, either commercially or noncommercially, provided that this License, the copyright notices, and the license notice saying this License applies to the Document are reproduced in all copies, and that you add no other conditions whatsoever to those of this License. You may not use technical measures to obstruct or control the reading or further copying of the copies you make or distribute. However, you may accept compensation in exchange for copies. If you distribute a large enough number of copies you must also follow the conditions in section 3.
You may also lend copies, under the same conditions stated above, and you may publicly display copies.

3. COPYING IN QUANTITY

If you publish printed copies (or copies in media that commonly have printed covers) of the Document, numbering more than 100, and the Document's license notice requires Cover Texts, you must enclose the copies in covers that carry, clearly and legibly, all these Cover Texts: Front-Cover Texts on the front cover, and Back-Cover Texts on the back cover. Both covers must also clearly and legibly identify you as the publisher of these copies. The front cover must present the full title with all words of the title equally prominent and visible. You may add other material on the covers in addition. Copying with changes limited to the covers, as long as they preserve the title of the Document and satisfy these conditions, can be treated as verbatim copying in other respects.
If the required texts for either cover are too voluminous to fit legibly, you should put the first ones listed (as many as fit reasonably) on the actual cover, and continue the rest onto adjacent pages.
If you publish or distribute Opaque copies of the Document numbering more than 100, you must either include a machine-readable Transparent copy along with each Opaque copy, or state in or with each Opaque copy a computer-network location from which the general network-using public has access to download using public-standard network protocols a complete Transparent copy of the Document, free of added material. If you use the latter option, you must take reasonably prudent steps, when you begin distribution of Opaque copies in quantity, to ensure that this Transparent copy will remain thus accessible at the stated location until at least one year after the last time you distribute an Opaque copy (directly or through your agents or retailers) of that edition to the public.
It is requested, but not required, that you contact the authors of the Document well before redistributing any large number of copies, to give them a chance to provide you with an updated version of the Document.

4. MODIFICATIONS

You may copy and distribute a Modified Version of the Document under the conditions of sections 2 and 3 above, provided that you release the Modified Version under precisely this License, with the Modified Version filling the role of the Document, thus licensing distribution and modification of the Modified Version to whoever possesses a copy of it. In addition, you must do these things in the Modified Version:

A. Use in the Title Page (and on the covers, if any) a title distinct from that of the Document, and from those of previous versions (which should, if there were any, be listed in the History section of the Document). You may use the same title as a previous version if the original publisher of that version gives permission.
B. List on the Title Page, as authors, one or more persons or entities responsible for authorship of the modifications in the Modified Version, together with at least five of the principal authors of the Document (all of its principal authors, if it has fewer than five), unless they release you from this requirement.
C. State on the Title page the name of the publisher of the Modified Version, as the publisher.
D. Preserve all the copyright notices of the Document.
E. Add an appropriate copyright notice for your modifications adjacent to the other copyright notices.
F. Include, immediately after the copyright notices, a license notice giving the public permission to use the Modified Version under the terms of this License, in the form shown in the Addendum below.
G. Preserve in that license notice the full lists of Invariant Sections and required Cover Texts given in the Document's license notice.
H. Include an unaltered copy of this License.
I. Preserve the section Entitled "History", Preserve its Title, and add to it an item stating at least the title, year, new authors, and publisher of the Modified Version as given on the Title Page. If there is no section Entitled "History" in the Document, create one stating the title, year, authors, and publisher of the Document as given on its Title Page, then add an item describing the Modified Version as stated in the previous sentence.
J. Preserve the network location, if any, given in the Document for public access to a Transparent copy of the Document, and likewise the network locations given in the Document for previous versions it was based on. These may be placed in the "History" section. You may omit a network location for a work that was published at least four years before the Document itself, or if the original publisher of the version it refers to gives permission.
K. For any section Entitled "Acknowledgements" or "Dedications", Preserve the Title of the section, and preserve in the section all the substance and tone of each of the contributor acknowledgements and/or dedications given therein.
L. Preserve all the Invariant Sections of the Document, unaltered in their text and in their titles. Section numbers or the equivalent are not considered part of the section titles.
M. Delete any section Entitled "Endorsements". Such a section may not be included in the Modified version.
N. Do not retitle any existing section to be Entitled "Endorsements" or to conflict in title with any Invariant Section.
O. Preserve any Warranty Disclaimers.

If the Modified Version includes new front-matter sections or appendices that qualify as Secondary Sections and contain no material copied from the Document, you may at your option designate some or all of these sections as invariant. To do this, add their titles to the list of Invariant Sections in the Modified Version's license notice. These titles must be distinct from any other section titles.
You may add a section Entitled "Endorsements", provided it contains nothing but endorsements of your Modified Version by various parties—for example, statements of peer review or that the text has been approved by an organization as the authoritative definition of a standard.
You may add a passage of up to five words as a Front-Cover Text, and a passage of up to 25 words as a Back-Cover Text, to the end of the list of Cover Texts in the Modified Version. Only one passage of Front-Cover Text and one of Back-Cover Text may be added by (or through arrangements made by) any one entity. If the Document already includes a cover text for the same cover, previously added by you or by arrangement made by the same entity you are acting on behalf of, you may not add another; but you may replace the old one, on explicit permission from the previous publisher that added the old one.
The author(s) and publisher(s) of the Document do not by this License give permission to use their names for publicity for or to assert or imply endorsement of any Modified Version.

5. COMBINING DOCUMENTS

You may combine the Document with other documents released under this License, under the terms defined in section 4 above for modified versions, provided that you include in the combination all of the Invariant Sections of all of the original documents, unmodified, and list them all as Invariant Sections of your combined work in its license notice, and that you preserve all their Warranty Disclaimers.
The combined work need only contain one copy of this License, and multiple identical Invariant Sections may be replaced with a single copy. If there are multiple Invariant Sections with the same name but different contents, make the title of each such section unique by adding at the end of it, in parentheses, the name of the original author or publisher of that section if known, or else a unique number. Make the same adjustment to the section titles in the list of Invariant Sections in the license notice of the combined work.
In the combination, you must combine any sections Entitled "History" in the various original documents, forming one section Entitled "History"; likewise combine any sections Entitled "Acknowledgements", and any sections Entitled "Dedications". You must delete all sections Entitled "Endorsements".

6. COLLECTIONS OF DOCUMENTS

You may make a collection consisting of the Document and other documents released under this License, and replace the individual copies of this License in the various documents with a single copy that is included in the collection, provided that you follow the rules of this License for verbatim copying of each of the documents in all other respects.
You may extract a single document from such a collection, and distribute it individually under this License, provided you insert a copy of this License into the extracted document, and follow this License in all other respects regarding verbatim copying of that document.

7. AGGREGATION WITH INDEPENDENT WORKS

A compilation of the Document or its derivatives with other separate and independent documents or works, in or on a volume of a storage or distribution medium, is called an "aggregate" if the copyright resulting from the compilation is not used to limit the legal rights of the compilation's users beyond what the individual works permit. When the Document is included in an aggregate, this License does not apply to the other works in the aggregate which are not themselves derivative works of the Document.

If the Cover Text requirement of section 3 is applicable to these copies of the Document, then if the Document is less than one half of the entire aggregate, the Document's Cover Texts may be placed on covers that bracket the Document within the aggregate, or the electronic equivalent of covers if the Document is in electronic form. Otherwise they must appear on printed covers that bracket the whole aggregate.

8. TRANSLATION

Translation is considered a kind of modification, so you may distribute translations of the Document under the terms of section 4. Replacing Invariant Sections with translations requires special permission from their copyright holders, but you may include translations of some or all Invariant Sections in addition to the original versions of these Invariant Sections. You may include a translation of this License, and all the license notices in the Document, and any Warranty Disclaimers, provided that you also include the original English version of this License and the original versions of those notices and disclaimers. In case of a disagreement between the translation and the original version of this License or a notice or disclaimer, the original version will prevail.

If a section in the Document is Entitled "Acknowledgements", "Dedications", or "History", the requirement (section 4) to Preserve its Title (section 1) will typically require changing the actual title.

9. TERMINATION

You may not copy, modify, sublicense, or distribute the Document except as expressly provided under this License. Any attempt otherwise to copy, modify, sublicense, or distribute it is void, and will automatically terminate your rights under this License.

However, if you cease all violation of this License, then your license from a particular copyright holder is reinstated (a) provisionally, unless and until the copyright holder explicitly and finally terminates your license, and (b) permanently, if the copyright holder fails to notify you of the violation by some reasonable means prior to 60 days after the cessation.

Moreover, your license from a particular copyright holder is reinstated permanently if the copyright holder notifies you of the violation by some reasonable means, this is the first time you have received notice of violation of this License (for any work) from that copyright holder, and you cure the violation prior to 30 days after your receipt of the notice.

Termination of your rights under this section does not terminate the licenses of parties who have received copies or rights from you under this License. If your rights have been terminated and not permanently reinstated, receipt of a copy of some or all of the same material does not give you any rights to use it.

10. FUTURE REVISIONS OF THIS LICENSE

The Free Software Foundation may publish new, revised versions of the GNU Free Documentation License from time to time. Such new versions will be similar in spirit to the present version, but may differ in detail to address new problems or concerns. See http://www.gnu.org/copyleft/.

Each version of the License is given a distinguishing version number. If the Document specifies that a particular numbered version of this License "or any later version" applies to it, you have the option of following the terms and conditions either of that specified version or of any later version that has been published (not as a draft) by the Free Software Foundation. If the Document does not specify a version number of this License, you may choose any version ever published (not as a draft) by the Free Software Foundation. If the Document specifies that a proxy can decide which future versions of this License can be used, that proxy's public statement of acceptance of a version permanently authorizes you to choose that version for the Document.

11. RELICENSING

"Massive Multiauthor Collaboration Site" (or "MMC Site") means any World Wide Web server that publishes copyrightable works and also provides prominent facilities for anybody to edit those works. A public wiki that anybody can edit is an example of such a server. A "Massive Multiauthor Collaboration" (or "MMC") contained in the site means any set of copyrightable works thus published on the MMC site.

"CC-BY-SA" means the Creative Commons Attribution-Share Alike 3.0 license published by Creative Commons Corporation, a not-for-profit corporation with a principal place of business in San Francisco, California, as well as future copyleft versions of that license published by that same organization.

"Incorporate" means to publish or republish a Document, in whole or in part, as part of another Document.

An MMC is "eligible for relicensing" if it is licensed under this License, and if all works that were first published under this License somewhere other than this MMC, and subsequently incorporated in whole or in part into the MMC, (1) had no cover texts or invariant sections, and (2) were thus incorporated prior to November 1, 2008.

The operator of an MMC Site may republish an MMC contained in the site under CC-BY-SA on the same site at any time before August 1, 2009, provided the MMC is eligible for relicensing.

How to use this License for your documents

To use this License in a document you have written, include a copy of the License in the document and put the following copyright and license notices just after the title page:

Copyright (c) YEAR YOUR NAME.

Permission is granted to copy, distribute and/or modify this document

under the terms of the GNU Free Documentation License, Version 1.3

or any later version published by the Free Software Foundation;

with no Invariant Sections, no Front-Cover Texts, and no Back-Cover Texts.

A copy of the license is included in the section entitled "GNU

Free Documentation License".

If you have Invariant Sections, Front-Cover Texts and Back-Cover Texts, replace the "with...Texts." line with this:

with the Invariant Sections being LIST THEIR TITLES, with the

Front-Cover Texts being LIST, and with the Back-Cover Texts being LIST.

If you have Invariant Sections without Cover Texts, or some other combination of the three, merge those two alternatives to suit the situation.

If your document contains nontrivial examples of program code, we recommend releasing these examples in parallel under your choice of free software license, such as the GNU General Public License, to permit their use in free software.